ECO-MANAGEMENT AND ECO-AUDITING:

Environmental Issues in Business

Compiled and edited by
Linda S Spedding, LLB (Hons), LLM, PhD
David M Jones
Christopher J Dering, MA

Chancery Law Publishing

A Division of John Wiley & Sons Ltd,
London, New York, Chichester, Brisbane, Toronto, Singapore.

Published in the United Kingdom by
Chancery Law Publishing Ltd
22 Eastcastle Street
London W1N 7PA

Published in North America by
John Wiley & Sons Inc
7222 Commerce Center Drive
Colorado Springs, CO 80919
USA

Typeset by
Vision Typesetting, Manchester

Printed in Great Britain by
Ipswich Book Company Ltd

ISBN 0471 93693 6

A copy of the CIP entry for this book is available
from the British Library

Library of Congress Cataloging-in-Publication Data

Eco-management and Eco-auditing: environmental issues in business /
edited by Linda S Spedding, David M Jones, Christopher J Dering
p. cm.
Includes bibliographical references and index.
ISBN 0–471–93693–6 (cloth)
1. Industrial management – Environmental aspects – Europe.
2. Environmental law – Europe. 3. Industrial management –
Environmental aspects – Europe – Auditing. 4. Environmental policy –
Europe. I. Spedding, Linda S, 1952– . II. Jones, David M.
III. Dering, Christopher J.
HD69.P6E27 1993
658.4′08–dc20 93–7446 CIP

Dedicated to the late
Lynette Romans Van Schaik-Jones
A World Traveller

Contents

CONTENTS

CONTENTS

Preface

A few months ago I was walking with my two daughters, Lisa and Karen, aged 14 and 11, through Palm Springs. The temperature was 120 degrees Fahrenheit but the grass was very green. It occurred to us that if the water tap to this delightful desert resort in California were to be switched off, there would be no Palm Springs and the natural environment, in the form of desert and scrub, would again take over. It would, of course, be impracticable simply to turn off the tap, thereby bringing the community to an end, but my daughters and I did wonder whether the world would ever build another Palm Springs. The answer is probably that it would not, given the growing awareness of the environment, population and energy resources, the increasing risk to the planet Earth and the effect of market forces.

Governments have responded to the change in awareness as to environmental issues with an enormous volume of legislation, particularly from the United States and Europe but this raises the question whether law and its enforcement is a suitable vehicle by which to implement changes in practice in this field. Can the law accommodate commercial considerations and companies' obligations to their shareholders? Is it sensible to send a company director to prison for a breach of environmental regulations?

The increasing emphasis on the role of environmental management systems and Eco-audits represents a start, albeit in their present voluntary form a small one, down the path towards encouraging management compliance rather than legal enforcement. This approach seeks to introduce management changes and to produce a different management approach to environmental issues. The law's interference in management systems is already becoming apparent to the construction industry in Europe with the arrival of the Construction Sites Directive from Brussels; its intention is to improve the comparatively poor health and safety record on construction sites through new mandatory management procedures. With the growing tide of public accountability and publicity in relation to performance deficiencies around the world, companies who fail to comply with these new management systems and requirements may find themselves omitted from tender lists, denied trading licences, refused project finance; individuals who fail properly to implement such systems may be debarred from becoming senior managers or directors. It is at the beginning rather than at the end of the transactional cycle that one is likely to experience the growing significance of the combination of management systems and the law.

This book is divided into ten chapters. The first three set the scene for the Eco-audit and the potential implications for an overall corporate policy. The next two review various management systems and standards both nationally and internationally followed by the inter-related topic of health and safety management. Chapters 7 and 8 look more closely at specific legal liability issues including the development of land and the problems of contamination. Chapter 9 deals with environmental risk protection including warranties and indemnities and the growing demand of insurance companies that proper environmental management systems and audits are in place before any consideration of an environmental risk. It also touches upon confidentiality. The last chapter moves forward in time to the increasing use of information technology as a tool to set up, monitor and review management systems as governed by the growing influence of regulatory law. Generally we have sought to draw the line for each chapter as at December 1992.

Proceeding from an EC/UK legal standpoint, the contributions which make up this book raise issues rather than seeking to answer specific problems. Its scope has been, of course, driven by the forthcoming EC Eco-audit legislation but reflects a wider range of matters of an environmental nature which are likely to be of importance to those involved in commerce generally. My personal belief is that it will represent a useful addition to the lively debate which presently surrounds the questions of how and why protection of the environment should affect business life and the extent to which social pressure and changing attitudes should be reflected by mandatory legal requirements rather than attempts to secure voluntary participation. The collection has been put together on the basis that each chapter could stand on its own or alternatively the book could be read as a whole. As such, we hope that it will accommodate the passenger on a flight from London to Brussels to visit the Commission or from London to Los Angeles . . . perhaps to visit Palm Springs.

David Jones
December 1992

Editors' acknowledgements

In compiling this volume, the editors have relied heavily on contributions from colleagues at Masons and friends outside. To all those who have assisted with this publication the editors wish to express their thanks. Particular contributions have been provided by Anne Buchanan and Mark Harnett (Chapter 4), Mark Lane (Chapters 1 and 4), Barbara Linehan (Chapter 1), Bonnie Martin, John Moritz and Lesley Webber (Chapter 7), Iain Monaghan (Chapters 3, 8 and 9), Richard Stewart (Chapter 1), Amanda Stubbs (Chapter 5) and Richard Susskind (Chapter 10), all of Masons; and Steve Halls of EML; Malcolm Hutton of ERL; Robert Holmes, Lynne Miller and Michael Murphy of Environmental Strategies; John Pagella of Montague Evans and Mark Day of EAG Limited; Andrew Thornton of Fenchurch Group; and Mohammed Zairi of The European Centre for Total Quality Management at the University of Bradford. Beth Lang and Susan Olton have provided invaluable assistance in handling secretarial matters. Bih Wong of Masons and Claire Gilman as our external consultant editor have eased the production process greatly. Finally our thanks for the encouragement and support of our publishers at Chancery.

CONTENTS OF CHAPTER 1

Environmental Issues in Business: an Overview

Chapter 1

Environmental Issues in Business: an Overview

The integration of economics and ecology

As the 1987 Report of the World Commission on Environment and Development (The Brundtland Report) states, until recently the world was a place in which human activities and their effects were neatly compartmentalised within nations, within sectors (energy, agriculture, trade etc), and within broad areas of concern such as environmental, economic and social. These compartments have dissolved.

The world is passing through a period of dramatic growth and fundamental change. Each step forward in growth itself triggers off a further and faster step forward. The existing world population of 5 billion will probably have to make room in a finite environment for approximately the same number again of human beings. The population of the world might stabilise at between 8 billion and 14 billion sometime in the next century according to UN projections. More than 90% of the increase is estimated to occur in the poorest countries and 90% of that growth in cities that are already overcrowded.

Economic activity has created a US$13 trillion world economy and it is estimated that this could grow 5- or 10-fold in the coming half century. Industrial production has grown more than 50-fold over the past century, four-fifths of this growth since 1950. These figures indicate profound impacts upon the environment, as humankind invests in houses, transport systems, farms and industries. Much of the economic growth draws on raw material from forests, soils, seas and waterways. A key factor of economic growth is new technology, and while this technology offers the potential for slowing the dangerously rapid consumption of finite resources, it also entails high risks including new forms of pollution. Meanwhile the industries most heavily reliant on environmental resources and most heavily polluting are growing most rapidly in the developing world, where there is both more urgency for growth and less capacity to minimise damaging side-effects.

These and other related changes have locked the global economy and global ecology together. We must concern ourselves with the impact of ecological stress – degradation of soils, water regimes, atmosphere and forests upon our economic prospects. Just as recently there has been a marked increase in economic interdependence amongst nations, we are now forced to face up to an accelerating ecological interdependence. Ecology and economy are becoming ever more interwoven, locally, regionally, nationally and globally into a seamless pattern of causes and effects (*cf. Our Common Future* (World Commission on Environment and Development, (Oxford University Press, 1987) 4–5).

Developing countries face the obvious life-threatening challenges of desertification, deforestation and pollution and endure most of the poverty associated with environmental degradation. All the countries of the world would suffer from the disappearance of rain forests in the tropics, and the consequent loss of plant and animal species, and changes in rainfall patterns. Industrial nations face the serious challenges of toxic chemicals, toxic wastes and acidification. All nations are likely to suffer from the releases by industrialised countries of carbon dioxide and of gases which react with the ozone layer (*ibid* 22).

National boundaries have become so porous that traditional distinctions between matters of local, national and international significance have become blurred. Ecosystems do not respect national boundaries. Water pollution moves through shared rivers, lakes and seas. The atmosphere carries air pollution over vast distances. Major accidents particularly those at nuclear reactors or at plants or warehouses containing toxic materials can have widespread regional effects.

Environmental issues transcend national boundaries, and failures to manage the environment and to sustain development threaten all countries. The environment and sustainable development are not separate challenges, they are inextricably linked.

The global dimension to the environmental issues that face us is recognised by the European Commission in its Fifth Environmental Action Programme published in March 1992 which states:

> "8. In the early stages, Community policy and action on the environment were mainly focused on the solution of particularly acute problems within the Community. Later there was a clearer recognition that pollution did not stop at its frontiers and that it was necessary, therefore, to intensify co-operation with third countries. In recent years, the evolution has gone a step further and it is now generally accepted that issues of a global nature – climate change, ozone depletion, diminution of biodiversity etc – are seriously threatening the ecological balance of our planet as a whole." (*Towards sustainability: A European Community Programme of Policy and Action in relation to the Environment and Sustainable Development*, Vol II p4. COM (92) final, 27.3.92).

It, therefore, follows that economics and ecology must be completely integrated in decision-making and law-making processes not just to protect the environment, but also to protect and promote development.

The challenges are both interdependent and integrated, requiring comprehensive approaches and participation at all levels. There is a growing need for effective international co-operation to manage ecological and economic interdependence. The rate of change is outstripping the ability of scientific disciplines and our current capabilities to assess and advise. The seriousness of this situation merits a preventative as opposed to a reactive approach. But even in the realms of a preventative approach, there are many key strategic variants. Debate over the use of regulatory fiscal and voluntary mechanisms, as well as their interaction, continues.

By way of illustration, discussion has taken place over the possible use of taxes, for example to reduce the emission of environmentally harmful gases, notably carbon dioxide, from various sources of energy. Taxation is clearly one means of achieving this result. However, there is a persuasive argument that market-based instruments will often be more efficient and less expensive than regulation in reducing emissions because they allow producers and consumers, rather than regulators, to decide how energy can best and most economically be used.

Further, long-term measures affecting the relative price of energy can only sensibly be taken when competitor countries are prepared to take similar action. Unilateral action by one country would do little to influence global warming. It would have a damaging impact on activity and employment in the energy intensive sectors of that country because it would make those sectors less competitive in the world market.

The threats, particularly that of global warming, facing the world's environment can be brought under control only through a world-wide co-operative effort. This is going to require a firm hand, vision and good faith from all nation states in the developed and developing worlds. Controlling these threats, and in particular that of global warming, could possibly be the first occasion on which the world adopts market-based solutions to major international problems of pollution control.

In industrialised nations, stabilising carbon dioxide emissions will require careful planning and substantial investment by industry, in particular the transport and energy industries and by vehicle manufacturers. No less importantly, it will require change, in our behaviour as individuals. Government can help to set the right framework for these changes but achieving reductions in carbon dioxide emissions will depend upon individual action. The self-discipline imposed by environmental management systems, such as those set out in BS 7750 and the evolving international standards, will have a key role to play in achieving such stabilisation of carbon dioxide emissions.

Technical aspects and concerns

As one of the more topical transboundary issues, it is useful to touch upon the problems of economic development in relation to air quality by way of illustration. The late 1980s and the early 1990s have seen international environmental agreements being increasingly introduced to control both global and regional air pollution problems. Examples include the Montreal Protocol to reduce the release of chloro-fluorocarbons (CFCs) and the Economic Commission for Europe (ECE) Trans-boundary Conventions on the emissions of NO_x and SO_x – the precursors of acid rain. More recently, the ECE Protocol on Volatile Organic Compounds (VOCs), designed to reduce the problem of ground level ozone formation, was signed in late 1991.

For many, particularly politicians, these agreements were hailed as successes and considered to symbolise a new spirit of co-operation between the nations of the world. However, the detailed workings of such agreements have been frustrated by the technical and economic difficulties of making the necessary changes. Many countries have simply been unable to make the required investment to comply with these arrangements. For example, many Central and Eastern European countries signed to ECE Conventions on NO_x and SO_x emissions. However, these countries have a redundant stock of power generating facilities which discharge large quantities of the acidic gases. Any economic recovery of Poland and the Czech and Slovak Republics can only lead to increases in pollutant emissions, in line with increased electricity demand. As such, the ability of these countries and others, to conform to the ECE conventions is problematic. More importantly, the serious problem of acid rain in the European region is likely to be exacerbated as long as existing power stations without controls operate in Central and Eastern Europe.

Scientists and policy makers working on acid rain in Europe now accept the need to focus on carefully devised and costed abatement strategies. These strategies take account of the variability in sensitivity to acid rain in different parts of Europe as well as the atmospheric transport of the pollutant emissions.

Currently, abatement strategies for acid rain in Europe are focusing on the actions needed to:

- minimise damage to sensitive areas;
- minimise the number of cases where critical loads of acid rain are exceeded; and
- minimise the area receiving acid rain.

These kinds of abatement strategies are likely to recommend that countries such as Germany and Sweden pay Central and Eastern European nations to build less polluting power stations rather than make further investments in their indigenous pollution control programmes.

The Montreal Protocol was hailed by many authorities as a landmark for global environmental management. This was partly because so many countries, over 70, were party to the Protocol, but also because governments viewed it as an agreement reached before the problem was, in their opinion, proved beyond doubt. The Protocol initially introduced a progressive reduction in the production and consumption of CFCs by the parties to the agreement. The Protocol acknowledges the existence of differing conditions around the world and has made provisions regarding developing countries and those which are members of economic organisations.

Since the original conditions of the Protocol came into force, it has been shown that the rate of ozone depletion in the upper atmosphere is increasing at twice the rate originally estimated by scientists. With this disturbing background, two further meetings of the parties to the Montreal Protocol have taken place – these were in January 1990 in London and in Copenhagen in October 1992. Both meetings introduced stricter controls on CFCs – in the London meeting developed countries agreed to phase out CFCs altogether by the year 2000. However, between the London and Copenhagen meetings disturbing evidence had accumulated pointing to record ozone depletion over Antarctica and increasing ozone deficiency in the Northern Hemisphere.

The Copenhagen meeting agreed to further accelerate the phase out of CFCs but controversially, also allowed major increases in the consumption of hydro-chloro-fluorocarbons (HCFCs). These chemicals have been developed as interim substitutes for CFCs, but they themselves are ozone depletors albeit much less powerful than CFCs. This allowance in effect provides a green light to both the manufacturers of HCFCs and to the producers of refrigerators to use these chemicals. There is now a concern that industry will generate many tonnes of HCFCs in the next decade despite scientific evidence that all forms of ozone depleting chemicals must be rapidly and effectively controlled.

To quote from the Declaration of ECO Asia 91 in which many environmental ministers and parliamentarians from Asia and the Pacific region participated:

> "The time for far reaching change has come. This change will entail creating sustainable social and economic systems on a global scale. Such a transformation will include a critical re-examination of social and economic activities on a scale that might be called revolutionary, encompassing the behaviour of industry, commerce and consumers, as well as the processes of production. Reform of our high consumption life styles would be essential. This, in its entirety, may be called the Eco Industrial Revolution." (*The Developing World and Emerging Economies – A view from Asra* by Ambassador Nobutoshi Akao, Ministry of Foreign Affairs, Japan at The *Sunday Times* International Conference on the Environment, July 1991.)

One facet of this Eco Industrial Revolution is the development of international standards of environmental management. A step on the way to a world in which environmental regulation has moved beyond the usual menu of safety regulations, zoning laws and pollution control enactments to a world where environmental objectives are built into prior approval procedures for investment and technology choice taxation, following trade incentives and all components of development policy.

The European Commission's Fifth Environmental Action Programme

In Volume 2 Chapter 7 of its Fifth Environmental Action Programme, the European Commission recognises that there will be a continuing need for legislative measures at Community level, particularly in respect of:

"– the establishment of fundamental levels of environmental care and protection;
– community commitments to wider international agreements; and
– common standards and/or controls which may be deemed necessary or expedient to preserve the integrity of the internal market."

(*Towards sustainability: A European Community Programme of Policy and Action in relation to the Environment and Sustainable Development* Vol II p 64. COM (92) final, 27.3.92.)

However, the Fifth Environmental Action Programme recognises that it is not feasible to adopt a Directive or Regulation which says: "Thou shalt act in a sustainable manner". It also recognises that because of the broad scope of many of the present-day environment issues and the threats to ecology posed by current trends in political, economic and social life, it is imperative to focus on the causes of environment-based problems in a different manner. It is essential to go to the root of these problems – human activity, human values in relation to the environment and natural resources and human behaviour and consumption patterns.

The Fifth Environmental Action Programme goes on to argue that in order to bring about substantial changes in current trends and practices and to involve all sectors of society in a spirit of shared responsibility, a broader mix of "instruments" needs to be developed and applied. It continues that environmental policy will rest on four main sets of instruments:

– Regulatory Instruments;
– Market Based Instruments (including economic and fiscal instruments and voluntary agreements);
– Horizontal Supporting Instruments (research, information, education etc); and
– Financial Support Mechanisms.

(*Ibid* 64.)

Within the category of "Market Based Instruments", the Fifth Environmental Action Programme recognises environmental auditing. This is seen in the first instance as an internal management tool which should indicate performance on resource management including use of raw materials, energy consumption, productivity levels and waste. It should highlight areas of risk, indicate performance on risk prevention, and management and audit levels of safety and health in the workplace. For shareholders, investors, financial and insurance institutions, such resource-based audits would provide a performance indicator as important as traditional financial accounts. For public authorities and the public at large, environmental audits will provide both a compliance and performance indicator which can provide a boost to overall public confidence. (*Ibid* 68.)

In considering these issues the remainder of this introductory chapter will explore three matters. First, a brief look at developments in Europe; secondly, developments worldwide. Thirdly, the question of terminology: is audit the right word?

The development of an environmental policy for Europe

In order to understand the role of Eco-audits in EC environmental policy one must consider briefly the development of that policy over the last two decades. This is a topic which will be treated in more detail in Chapter 2. It is also necessary to examine some of the basic problems in using the law as a means of implementing environmental policy. In order to achieve improvement to the environment a balance must be made between regulatory and other, such as economic, intervention.

Most people in the developed world agree on basic standards of behaviour which are essential to the preservation of civilised society. Murder, theft and rape are all beyond the pale. These are, as such, the subject of state intervention as, on a less serious level, are "offences" such as negligence. This intervention can take the form of criminal sanctions, or the law can lay down rules for compensating victims of a particular type of conduct.

In law, most victims are self-defining. Dead bodies and outraged, burgled home owners are readily identifiable. In civil law, if the plaintiff can prove he has suffered loss, then he is entitled to compensation, whether the actions of the defendant are morally reprehensible or not. The problem for lawmakers comes when either there is no consensus that a form of conduct should be regarded as unlawful, or when victims are not easily identifiable. Environmental issues often cause difficulty on both these counts.

Environmental law is developing against a background where it has become increasingly difficult to identify victims. To look at the problem in another way, there have been so many victims that the principles of causation which once underpinned both criminal and civil liability are no longer able to cope with the potential for damage inherent in industrial progress. At the same time the public have begun to see themselves and future generations as potential victims of environmentally irresponsible behaviour.

One has to see current environmental legislation in the same light as civil rights laws in the United States and race discrimination laws in Europe. It is about punishing what is regarded as anti-social behaviour, but primarily it is about changing attitudes. There are different views about the use of criminal law as an instrument of social change: arguably, the legal regime created to combat racial discrimination will fail to eliminate prejudice, because not enough has been done to deal with the underlying causes of such prejudice. Certainly, environmental legislation will fail to solve the problems of achieving the correct balance between economic progress and ecologically responsible behaviour, unless both businesses and consumers can be convinced of the necessity and sense in changing many of their current attitudes.

The European Community, in its environmental policy, has adopted a mixture of coercion and self-regulation. As is made clear in the EC Fifth Environmental Action Programme, the focus of which is "sustainable development", key elements of the EC environmental policy include the use of market forces to improve the environment, with coercion and enforcement being called upon as a last resort. One could say that this is a somewhat artificial distinction, given that governments control the economic climate, so that market forces are actually a form of indirect coercion. This is especially true if those forces operate against a legislative framework which ultimately relies on the sanctions of criminal law, allows specific classes of civil action and creates new rights of access to information in order to enforce a social policy.

It is suggested, however, that the EC approach is the right one: the alternative to what is essentially a regime of self-regulation is wholesale and rigorous enforcement. This could include mandatory audits, but it would also require a system of severe penalties to ensure that companies did not simply cover themselves by building into their costs the fines imposed for polluting the environment. Such penalties have been imposed both in the United States and Europe, but generally on the basis that the offence was of such magnitude, or that the defendant had been given sufficient opportunity to, literally, clean up its act, that there was no alternative.

If such penalties were to become the norm rather than the exception, then the basic issue of sustainable development would be called into question. Whilst the insurance industry might bear the brunt for a short period, one would inevitably reach the position that a company could quite simply go out of business for one environmental misdemeanour. The social and economic consequences of that would be insupportable, and unacceptable to all but the most verdant of environmentalists.

Public awareness in Europe about environmental issues has undergone an immense change over the last 20 years. It still has a considerable way to go and, looking to the United States, Europe's law makers and enforcers will need, especially in a time of recession, to ensure that they do not run too far ahead of public opinion and produce anti-environmental backlash.

The Fifth Environmental Action Programme recognises this point stating that: "sustainable development implies putting in place a policy and strategy for continued economic and social development without detriment to the environment and the natural resources on which human activity depends."

Among other things, "sustainable development" implies that:

- the environment and its natural resources must be recognised as the basic foundation of all human activity and their satisfactory guardianship is a precondition for sustainable development;
- the flow of substances from the raw material stage through processing and production and on to consumption and waste should, as far as possible, become cyclic;
- production and consumption of energy should be rationalised; and
- behavioural trends in society itself should be altered.

In the Fifth Environmental Action Programme, the Commission is careful to emphasise the positive side to the new "sustainable" industrial policy:

- the new opportunities for business in "clean" technologies;
- the chance to attract new customers with more environmentally attractive products; and

– the chance to economise on the use of raw materials, energy and waste disposal.

It is suggested that some of the costs involved in improving the environmental performance of a business will be offset by substantial savings and exploitation of the new opportunities listed above.

The advent of the Eco-audit

The shift towards a new approach in environmental regulations has been shown by the establishment of a new legislative framework for environmental management, of which the Eco-audit forms a part. This approach abandons the traditional segmented approach whereby environmental legislation was divided according to the various media (*e.g.* sea, air, land, water, soil and wildlife). Rather than depending on regulation, it stresses the importance of information, awareness and responsibility, as well as the role of market mechanisms in ensuring that high standards of environmental protection are reached.

A first and important element in this new framework was the Council Directive on the Environmental Impact of Certain Private and Public Development Projects (Council Directive 85/337/EEC of 27 June 1985, OJ (1985) L 175/40). This requires that an environmental impact assessment is carried out prior to the authorisation of certain categories of developments. The information on the environmental effects of a project must "be taken into consideration in the development consent procedure", the policy assumption being that if the procedures in the Directive are followed, development decisions will be made in a more environmentally conscious manner.

Importantly, the Directive includes guidelines for public participation in the consent procedure, consultation with parties affected by the project, and the conduct of the environmental assessment. This topic is dealt with in further detail in Chapter 7.

On 7 June 1990, the Council adopted a Directive on Freedom of Access to Information (Council Directive 90/313/EEC of 7 June 1990, OJ (1990) L 158/56). This is designed to ensure public access to and dissemination of information on the environment which is in the hands of public authorities. The principle of freedom of access to information is new to most Member States' legal systems. Historically, Member States' administrations favour secrecy over public right of access to official information. Some Member States have a limited tradition of rights of access to environmental information, coupled in some cases with rights of participation in the regulatory regime. The UK Control of Pollution Act of 1974 is a good example of such laws. The provision of such rights, even where they do exist, is haphazard and inconsistent (for further comment, see Chapters 4 and 9).

The Directive has been widely criticised by environmental groups. There are several broadly-drafted exceptions to the general principle of freedom of access to environmental information. For example, there has been much controversy about the exception for "commercial and industrial confidentiality", terms which are undefined in the Directive. Nevertheless, refusals to supply information must be justified, and all refusals are subject to administrative or judicial review. While the Directive had to be implemented in all Member States by 31 December 1992, the Commission anticipated that its implementation would be difficult, given the breadth of the language used and the lack of definition of essential terms.

On 7 May 1990, the Council of Ministers adopted Regulation (EEC) 1210/90, which provides for the establishment of a European Environment Agency (EEA) and an Environmental Information and Observation Network. An important feature of the Agency is that participation will be open to non-EC members, such as the countries of the European Free Trade Association (EFTA) and Eastern European countries.

Even though, initially, the EEA will have no regulatory or enforcement powers, Commission officials have said that it should play a significant role in the formulation of environmental policy in the European Community. Achievement of this goal will be aided by the collection and publication of a large body of comparable data on the European environment, and the EEA's first task is to establish how to provide that data. To this end, the EEA and Member States will have to establish an information network on the environment throughout the Community.

The EEA's role will be reviewed after two years in operation. At that time, additional powers may be given to the EEA such as monitoring the implementation of EC environmental legislation in the Member States, and preparing eco-labels and criteria for the award of such labels to products, technologies and services which benefit the environment. The Regulation specifies in Article 21 that the EEA will begin operating once its venue has been decided. This decision on the location of the EEA has been blocked up to this point because of its linkage to negotiations regarding permanent seats of EC institutions, including the European Parliament.

The Council adopted Regulation (EEC) 880/92 on a Community eco-label award scheme on 23 March 1992 which sets uniform standards throughout the European Community for the issuing of eco-labels (OJ 1992 L99/1). The object of this initiative is to provide consumers with better information on the environmental impact of products and to promote those products which have lesser impact. As is further discussed in Chapter 4, the award of an eco-label will be based on a "cradle to grave" assessment of the product in question (*i.e.* from its production to its eventual disposal).

Another initiative which is relevant to the discussion of increased environmental awareness is in the field of packaging. The Commission adopted on 24 August 1992 a Directive on packaging and packaging waste for proposal to the Council. This attempts to incorporate the principle of the EC strategy on waste management into the field of packaging waste, and provides a framework for a harmonised EC policy on packaging and packaging waste. The objects of the draft are: (A) to harmonise national provisions in this area; (B) to prevent the creation of packaging waste; (C) to promote the recovery of non-avoidable packaging waste. Packaging is defined in the draft as:

> "all products made of any materials of any nature to be used for the containment, handling, delivery and presentation of goods, from raw materials to processed goods, from the producer to the user or the consumer. Disposables used for the same purpose are to be considered as packaging too."

The proposal applies to all packaging and packaging waste, whatever its use and composition, with the exception of packaging waste that qualifies as "hazardous" under other EC legislation. The draft sets quantified targets for recovery, minimisation and final disposal of packaging waste to be attained within 10 years. Within this period, Member States would also be obliged to reach specified intermediate targets.

Much work continues to be done by the Commission in the area of waste and liability for environmental damage. In 1989, it submitted a proposal to the Council for a Directive on civil liability for damage caused by waste which was based on the principle that the polluter must pay. Thus liability for damage to persons or to the environment caused by waste would be the responsibility of the producer of the waste, who would be strictly liable from the moment the waste was produced until it was lawfully transferred to a licensed waste disposal installation. The Commission produced an amended version of the proposal in 1991.

Work on this controversial initiative was stopped in 1991 while the Commission examined the broader problem of liability for damage to the environment, however caused. In a Green Paper the Commission advocates the establishment of a system of strict liability where the costs of restoration are met by an identifiable party. Where perpetrators cannot be identified, joint compensation funds are established to spread the costs of restorative action among specific economic sectors or geographical regions. This is discussed further in the context of contaminated land in Chapter 7.

The Council of Europe is in the final stage of drafting a convention on civil liability for environmental damage. The convention should be opened for signature next year. Under the convention, operators of activities classified as "dangerous" will be held liable for any damage caused to the environment. It is as yet unclear whether the European Community will base its legislation in this area on the convention.

Meanwhile, and bearing in mind the other EC developments, in 1990 the Commission made it clear that it was in the process of drafting legislation on environmental audits. Initial Commission papers outlined a plan for an ambitious scheme for the mandatory auditing of industrial facilities. Organisations consulted by the Commission during the course of its preparation showed considerable alarm at the extent of the plans. In particular, industry objected to the administrative and financial burdens which implementation of the proposals would entail. The scheme was increasingly unpopular given the recessionary climate so the Commission came up with a voluntary alternative which was eventually proposed to the Council on 18 December 1991. This is outlined below.

As will be seen in greater detail in Chapter 2, within the framework of the overall environmental policy, the Eco-audit aims to give companies a structure within which to:

- assess the impact of their environmental performance;
- establish an effective environmental protection scheme;
- evaluate environmental performance;
- stimulate developments; and
- inform the public about the environmental performance of the company.

Other world developments

Although the proposed EC Regulation on Eco-auditing is the first attempt to set out a comprehensive scheme for an environmental review and periodic audit, several countries and organisations both inside and outside the European Community have considered schemes along similar lines.

In France, for example, about 50,000 plants need a prefectorial authorisation in order to stay in business. These authorisations are issued after an inspection to establish that plants conform with legislation concerning emissions, waste management and health and safety. In addition, the Ministry of the Environment has targeted over 1,000 larger enterprises where internal monitoring systems should be installed to look at emissions, waste management and recycling on a systematic basis. As yet, however, there is no obligation to make the results of these studies public.

In the Netherlands, the government in 1989 required industries having a significant impact on the environment to develop and implement internal systems, including internal audits, and environmental policies with a view to improving their environmental performance. Eventually, this should affect over 30,000 establishments.

In Ireland, the newly formed Environmental Protection Agency is responsible for developing the necessary criteria for effective Eco-audits. The agency is also responsible for raising the profile of environmental issues with industry, especially in small- and medium-sized enterprises.

Sweden requires over 6,000 enterprises to submit an annual environmental report. The 600 largest have these reports independently verified and the results are available to the public. At the same time, the Swedish Confederation of Industry is encouraging its members to initiate a voluntary audit scheme.

The United States, however, has probably gone further than any other nation outside the European Community to encourage Eco-audits. In July 1986, the US Environmental Protection Agency (EPA) issued an Environmental Auditing Policy Statement which was designed to encourage the use of Eco-audits by US companies, and to lay down guide-lines for such audits. The EPA defined an Eco-audit as,

> ". . . a systematic, documented, periodic and objective review by a regulated entity of facility operations and practices related to meeting environmental requirements."

When, in 1988, the International Chamber of Commerce (ICC) issued its *ICC Guide to Effective Environmental Auditing*, it defined an audit as:

> "A management tool comprising a systematic, documented, periodic and objective evaluation of the organisation, management and equipment are performing with the aim of contributing to safeguarding the environment . . ."

This was the definition adopted by the EC Commission when drafting the Community's Eco-audit Regulation. One phrase used by the EC definition which gives a significant clue as to the philosophy behind the scheme is "management tool".

More or less in parallel with developments at the EC level, the British Standards Institute (BSI) has been making its own contribution to the development of finalised standards for management of environmental matters. BS 5750, the British Standard on Quality Systems is well-known both in the United Kingdom and abroad – where it has been designated CEN 29000 in the European context of the CEN/CENELEC rules and has much to do with the ISO 9000 standards.

More recently, the British Standard Specification for Environmental Management Systems has been published. BS 7750 has come into effect against the background that has been referred to earlier of ecology and economy becoming ever more interwoven locally, regionally, nationally and globally into a seamless pattern of causes and effects.

For this reason and also to maintain and increase their market share, organisations of all kinds are becoming increasingly concerned to achieve and to be able to demonstrate sound environmental practice. This concern has also to be seen in the context of increasingly stringent legislation (such as the UK Environmental Protection Act 1990) and the development of economic and fiscal measures to foster environmental protection. Accordingly, many organisations are now undertaking environmental reviews or audits to assess their environmental performance.

On their own, however, such "reviews" and "audits" cannot provide an organisation with the assurance that its performance not only meets but will continue to meet legislative and policy requirements. They are after all nothing more than a "snapshot in time". To be effective, such reviews need to be conducted within a structured management system, integrated with overall management activity and addressing all relevant aspects of environmental performance.

A system such as that envisaged in BS 7750 enables an organisation to establish procedures to set an environmental policy and objectives, achieve compliance with them, and demonstrate such compliance to others. It should be added that the Standard is also intended to support certification schemes. Environmental management audits and environmental management reviews are inherent, but separate, parts of the system envisaged by BS 7750. Audits are to assess both the effectiveness of the environmental management system and the achievement of the environmental objectives. Reviews check the continuing relevance of the environmental policy, update the evaluation of environmental effects, and check the efficacy of audits and of follow up actions (BS 7750: (1992) 3).

BS 7750 is complementary to BS 5750 in that each standard takes parallel approaches to achieving and demonstrating compliance with specified requirements.

Further, BS 7750 complements (albeit that there are differences) the draft European Community Eco-audit Regulation. BSI intends that BS 7750 should be compatible with the European Commission's Regulation on Eco-auditing. The Commission itself has recognised the relative ease of BS 7750 being transferred into a European standard. The adoption of BS 5750 as a European standard creates a useful precedent for BS 7750. This suitability for adoption as a European standard is enhanced by the complementarity between BS 7750 and the draft Eco-audit Regulation.

Looking further afield it is interesting to note the initiative in India in a recent amendment to the Companies Act to include an environmental audit as an annual requirement.

Is "audit" the right word?

The word "audit" has appeared already numerous times in this chapter, but is it the correct word to describe the evaluation of a company's environmental performance? In the United States and elsewhere terms such as assessment, survey, surveillance, review and appraisal are in frequent use. The question is raised not for the sake of semantics, but because implicit in the choice of the word "audit" is the association with financial audits.

One common element of the different schemes proposed by the EPA, ICC and the European Community is flexibility. They all set out basic guidelines and particular areas which should be examined. All the schemes require a systematic methodology, but they leave the detailed choice of method to the auditor (or verifier) and his client. This is on the basis that audits are voluntary and are not a substitute for compliance with environmental legislation or inspections by the relevant regulatory agencies.

Financial audits, on the other hand, are mandatory in all the Member States of the European Community, and are an integral part of ensuring compliance with the regulation of commercial affairs. Whilst, at the end of the day, the aim of a financial audit is to give a true and fair picture of the financial state of affairs of a company, the auditors do not have a great deal of discretion as to how they go about their task. Certain matters are laid down by law, others are imposed by their professional bodies.

In addition, whilst auditors now invariably make suggestions about how a company can improve its financial systems and management, it is not the financial auditor's function to make recommendations about how a company might improve its manufacturing or marketing systems. For that, an auditor will call upon one of his colleagues in management consultancy. This highlights the major differences between a financial and Eco-audit. The former is primarily compliance oriented, the latter is performance based, and concerned with improving performance.

As is discussed elsewhere in this book, there are moves towards recognised standards nationally, European-wide and internationally, for environmental auditing. In the United States, the EPA publishes reviews of current practices in environmental auditing. The general view is that creditable Eco-auditing will be based on generally recognised methodology rather than an exhaustive list of environmental requirements. That has not, however, prevented the accountancy profession from studying the possibility of developing "green" accountancy standards. The Institute of Chartered Accountants in England and Wales has stated that the environment is increasingly becoming a cost item in companies' affairs and should be subject to the conventional accountancy framework.

There is still a long way to go, however, before this can be sensibly introduced. One can see the sense of incorporating certain environmental items into a balance sheet, such as contingent liabilities. Indeed, future clean-up costs and other legal liabilities should already be included in a company's accounts in many countries. More ambitious ideas, however, based on the concept of "sustainable development" and involving a distinction between "natural" and "man made" assets would seem problematic.

In the following chapters, various aspects of the EC Eco-audit Regulation and the problems which it seems likely to raise, both from a legal and a commercial point of view, are touched upon. At this relatively early stage in the lengthy and wide ranging debate which the Regulation seems set to foster, the aim of this book is not to seek so much to offer definitive answers and solutions, but rather to highlight areas of particular interest and to stimulate further discussion.

CONTENTS OF CHAPTER 2

The Eco-auditing Regulation

Chapter 2
The Eco-auditing Regulation

Preliminary discussions

The preliminary discussions relating to the proposal for an "Eco-audit" centred on a way of providing a compulsory, Community-based, regulatory framework for environmental auditing in enterprises. The initial work by the EC Commission was in fact for an environmental auditing Directive as opposed to a Regulation. This was commenced in June of 1990 and an early discussion document was circulated in the Autumn of that year. A more considered Consultation Paper was drafted in December 1990, which was then discussed at the first meeting of National Experts on the issue in January 1991. The "Draft Elements" of the proposed forthcoming Directive contained three principal elements:

(i) companies engaged in specified industrial activities would have to carry out a periodic self-assessment of their environmental performance;
(ii) the assessments carried out by an upper tier of these businesses would be subject to external verification by independent registered auditors; and
(iii) some of the findings would be reported to the public in the form of a periodic environmental statement.

Whereas some of the large European companies engaged in environmentally sensitive activities had already begun to carry out their own audits – that is, systematic examinations of the environmental effects of their operations – the extent of the initial Commission proposals met with some alarm when their 17-page Consultation Paper was circulated to industry associations. The environmental adviser to the UK Chemical Industry Association, estimating that the requirements would apply to well over 12,000 sites in Britain alone, stated that "the cost will be well into the millions of pounds each year". Moreover, tens of thousands of industrial sites throughout Europe would have to publish annual audits of their environmental performance according to the terms of the proposals. Indeed, the Commission listed 58 individual industries for which audits would be requested. These industries included manufacturers of chemicals, pharmaceuticals, plastics, cement, glass, detergents, artificial fibres, motor vehicles, aircraft, paper and board, dairy products, sweets, drinks and data processing equipment, as well as oil refining, electricity supply and waste disposal. The Commission required every site in more than half of these industries to carry out the audits: also the thresholds set for other industries were generally low, for instance, a factory employing over 25 people, manufacturing vegetable and animal fats would be covered.

The consultation document proposed a scheme whereby all the sites would have to carry out an environmental audit each year and that an annual statement would have to be published which would be inspected by the public and relevant authorities. The Commission also produced detailed annexes laying down some of the items to be covered by the audit, including "pollution discharge control, monitoring and reducing", "energy choice and reduction of energy use", "waste transportation, elimination, recycling and re-use", and "product planning (design, packaging, transportation, use and elimination)". A separate annex described a slimmed down statement which would be made available to the public and would include a "list of areas where the companies had difficulty in complying with environmental legislation". The Commission suggested that shareholders should be made aware of "long-term cost benefit analyses" of investments proposed to remedy defects revealed by the audit.

Following criticisms of the proposals, largely relating to the financial and administrative burden of compulsory audits, the threats to commercial confidentiality and the lack of enough experts to carry out such audits, the Commission re-drafted its proposals on the basis that the participation of companies in the scheme would be voluntary. This new approach occurred also because several polls conducted in the Member States showed a preference for the application of a voluntary system. Those volunteering to participate in the scheme would, of course, have to respect its provisions in order, for example, to make it of value. Moreover, the later drafts were in the form of a Regulation rather than a Directive and included a provision for review of the scheme after four years leaving open the possibility of it becoming mandatory in the future.

Interested parties

The usual EC legislative procedure provides for a consultation process between interested parties and the Commission in respect of relevant proposals. As part of this process, the Consultation Papers are circulated to national representative bodies, the relevant representatives of industry, trade associations and the like, for feedback which can, in turn, be presented where useful, to their European counterparts in Brussels. For example, in the United Kingdom the Department of the Environment (DoE) and the Department of Trade and Industry (DTI) arranged for the circulation of consultation documents and position papers, as well as expert meetings. Interestingly enough, due to the confusion over the term "audit" some of the discussion was based upon whether the title "environmental management" should replace that of "environmental audit". This was a view that the DTI and DOE shared and, indeed, the French representatives proposed this at an experts' meeting. The DTI and DOE believed that companies needed to integrate the environment in all of their activities. As Michael Massey of the DTI stated at a conference in London entitled "The European Commission's Proposals for Environmental Audits" held on 17 April 1991:

> "That means undertaking the sort of preliminary assessment which some people call audit to get a very good handle on the impact of their activities on the

environment, to develop a clear policy within the company, to ensure that everyone understands that policy, to have programmes to implement policy and arrangements to monitor their activities, and to audit the way in which the policy and the programmes are working in practice. We think it is better to regard all those, most of which are covered, including the last item on my list, which is reporting – to the shareholders, to the public, to the workforce, to the outside world; all of those are picked up in the European proposals – as environmental management, with the auditing label attaching only to the check of how the system is working."

Although in the event the European proposals have proceeded on the basis of the title "audit" – largely because of the influence of the practice of North American companies – the discussion remains valid. Also, the outcome of the consultation process has meant that a more clear-cut, well defined, scheme has emerged. Again, the recently published proposal has been circulated for comment to the interested parties prior to the discussion in the Council of Ministers, the intention is that the Regulation will be passed by the end of December 1992 and the scheme operable 18 months later, that is, mid-1994. (For the up-to-date time-table and relevant revised comment see the Epilogue.)

Moving objectives and changing drafts

As a result of this consultative approach, the Commission revised their proposal quite extensively, generally to make it much more flexible and to "loosen it up". For instance, whereas the first draft had a list of processes which would determine which companies would be required to undertake audits, the second proposal contained a slightly different list, suggesting what sort of companies should voluntarily undertake audits. In turn, that list disappeared. If the objective was to be the overall improvement of environmental performance then the scheme should be attractive to smaller organisations. Although existing auditing standards were relevant, for example, the guidelines that the International Chamber of Commerce (ICC) had published, these applied more readily to large multi-site companies rather than to smaller or medium-sized companies. If the environmental performance of companies as a whole was in fact to be improved, it was just as important for smaller companies to use such tools and techniques to improve their performance as it was for the very large companies. With this in mind, the earlier, rather ambitious and prescriptive approach of the Commission was softened.

The fact that the Commission also re-worked its proposals to make the scheme voluntary represented an important shift which may not, in the end, be final. The DTI, for example, believed that a voluntary approach was better because it enabled the encouragement of a commitment to environmental practices rather than the defensive legalistic response created by regulatory requirements. The DTI considered that the potential and actual benefits that many firms would find in using techniques of environmental auditing would be better obtained if the proposal did not *require* the undertaking of the activity with a report on it. Similarly, in keeping with the moving objectives, the initial proposals on regulating the so-called environmental auditing profession, the verifiers, were discussed in the changing drafts. Throughout the process the Commission was open to ideas and comments, not just from Governments, but also from industry bodies, environmental bodies and others who had a reasoned view to put to them. Presumably a lot will depend on the progress made under this voluntary scheme as to whether or not in the future there will be a mandatory requirement, how it is implemented and whether its coverage of companies will be defined by size or polluting potential.

The final Commission proposal

The Regulation that has evolved as a result of the earlier discussions and drafts was, as stated earlier, published in March 1992. As noted, the final instrument is a Regulation, rather than a Directive, which means that Member States have to have the system in place by the due date without the need for any national implementing legislation. The type of audit within the scheme has now become known as an Eco-audit.

In its final version, the Regulation establishes a Community environmental auditing scheme for certain industrial activities. The objective is to provide some incentive to the use of environmental auditing as a tool for systematic, periodic and objective evaluation of environmental performance. This evaluation covers certain industrial activities with the aim of facilitating effective management control by companies of their practices relevant to the environment, improving company authority over and public awareness of the environmental performance of these activities, and identifying the scope and ways for improving, as well as establishing an appropriate basis for company policy decisions aiming at improvements.

Within the scheme the audit is formally referred to as an "environmental audit" and is defined as:

> ". . . a management tool comprising a systematic, documented, periodic and objective evaluation of the performance of the organisation, management system and equipment designed to protect the environment with the aim of:
> (i) facilitating management control of the environmental practices;
> (ii) assessing compliance with company policies including observance of existing regulatory requirements."

The definition is taken almost verbatim from a definition first proposed by the ICC as quoted in Chapter 1, but the audit, as defined, does not stand alone. It fits within a novel comprehensive scheme commonly referred to as the Eco-audit scheme which is not found in any EC country. Article 1 of the final version (OJ 1992 C76/02) deals with the objectives of the Eco-audit scheme and provides:

> "1. A Community scheme allowing voluntary participation by industrial companies, hereinafter referred to as the 'Eco-audit scheme', is hereby estab-

lished for the evaluation and improvement of the environmental performance of industrial activities and the provision of the relevant information to the public.
2. The objective of the Eco-audit scheme shall be to promote improvements in the environmental performance of industrial activities by:
(a) the establishment and implementation of environmental protection systems by companies;
(b) the systematic, objective and periodic evaluation of the environmental performance of such systems;
(c) the provision of information on environmental performance to the public."

This represents the first time that the Commission has proposed environmental legislation not already in existence in one or more EC Member States. The details and terms of the legislation should be understood, therefore, by companies considering participation and are, therefore, assessed below. Although currently known as an Eco-audit, it should be understood that an "environmental audit" and an "Eco-audit" generally refer to similar issues. Generally in evaluating the interaction between any business operation and its surroundings, the following matters for example, would be examined:

(i) emissions to air, land and water;
(ii) legal requirements and constraints;
(iii) the effects on neighbouring communities, landscapes and ecology; and
(iv) public perception of the operating company.

A particular initiative is that the audit procedures and an environmental statement have to be verified by an external independent auditor under the Eco-audit scheme. The main audit stages (which are described further below) can therefore be outlined as follows:
(i) the planning of the audit activities and the defining of responsibilities;
(ii) the review of the environmental protection policy of the company;
(iii) an assessment of the organisation, management and equipment of a company;
(iv) the gathering of relevant data;
(v) the evaluation of overall performance;
(vi) the identification of areas for improvement; and
(vii) the internal reporting to top management.

Key features of an audit

It is suggested that whatever name one gives to this type of exercise, in order to be credible, it must exhibit certain easily identifiable features.

Systematic

"Systematic" has, in this context, two aspects: the methodology to be adopted and the scope of the audit. As regards methodology, it is interesting that the US Environmental Protection Agency (EPA), ICC and the European

Community all arrive at broadly similar views as to the organisation of an Eco-audit, which emphasise the importance of:

- management understanding which leads to support and commitment;
- a proper definition of the scope and aims of the audit;
- proper and adequate organisation and staffing;
- independence and objectivity of the audit team; and
- quality control.

As must be clear from the list above, all of these are interdependent. One cannot hope to arrive at management understanding and commitment without there first being a clear statement of the objectives behind an audit. Equally, without that commitment, it will be impossible to put in place the necessary organisation and staffing, and without adequate quality controls, it will be impossible to maintain and gauge the independence and objectivity of the audit team. Good documentation is also essential if the exercise is to be useful and credible.

As to the scope of an Eco-audit, one is in part looking at certain identifiable areas such as air emissions, water pollution, waste disposal and noise. These may be described as "compliance" items as the purpose of the audit is simply to check a company's adherence to statutory requirements and avoid the risk of legal proceedings. However, most Eco-audit schemes go beyond mere compliance with the law and expect valid audits also to examine areas such as efficient use and recycling.

Even the EPA, whose definition of an audit is, on the face of it, limited to compliance, makes it clear in its policy statement that one of the aims of an audit is to identify and cure potential environmental problems. Inevitably, this means that any audit programme will need to go beyond the check-list of relevant environmental legislation and take into account issues such as health and safety and employment policy. All of the schemes, therefore, demand a comprehensive statement from management at the highest level on their company's environmental policy. Under the EC scheme this can be a general declaration, a series of specific objectives, or a combination of the two. This is not an easy task.

The EC scheme, for example, includes, as one of the objectives of an Eco-audit, compliance with company environmental policy. Management, therefore, needs to resist the temptation to issue a glowing declaration of intent before an audit, in case it is over-ambitious and results in an unfavourable auditors' report, quite apart from the risk that such statements may increase the likelihood of successful civil liability claims.

Equally important is the definition of the scope of the audit in terms of geography, activities and organisation. The EC scheme approaches the problem by making the audit site-specific. There is sense in this approach. It takes into account varying factors which may prevail in different parts of a company's activities. It also emphasises the personal responsibility of management at site level which is essential for a proper implementation of environmental policy. Nevertheless, an effective audit will need at some point to pull together all of the site audits to produce an overall picture of an organisation's environmental performance, sometimes called the "holistic" or "cradle to grave" approach or, in the United States and Japan, "Total Quality".

Periodic

All the schemes require that audits be carried out at regular intervals. The EC scheme requires audits at one to three year intervals depending on the level of environmental impact of a site, which in turn is determined by criteria such as level of environmental concern, size and complexity of operations, volume of emissions, and history of environmental problems.

Objective

The basic principle is of having one's auditors approach their task unbiased. As the EPA states the avoidance of personal relationships, financial or other conflicts of interest, interference with free enquiry or judgment, or fear of potential retribution is essential to producing a credible audit. It also begs the question as to who the auditors should be.

On one level, there is a perfectly good case for a company having internal environmental auditors in the same way as many corporations have internal financial controls. However, just as internal financial auditors cannot produce their company's end of year accounts, so one needs external environmental auditors to produce Eco-audits.

The Eco-audit scheme

Article 2 sets out the key definitions for the purposes of the Regulation. In effect the Eco-audit Regulation contains a bewildering variety of defined terms which could be applied to a company's policy. A company is expected to have:

(i) an environmental protection system;
(ii) an environmental policy;
(iii) an environmental management system;
(iv) environmental objectives; and
(v) an environmental programme; and
(vi) it is expected to undertake environmental reviews and, of course, environmental audits and to issue environmental statements.

What is the relationship between all these terms?

One can divide the defined terms into four categories: those that deal with policy; those that deal with the control systems adopted to implement policy and internally monitor that implementation at a company-wide level; those that deal with the formulation of policy for a specific site, and its implementation; and those that deal with the external auditing of the company's policies.

The "environmental protection system" is at the heart of the scheme. It includes, in writing: an environmental policy, environmental objectives, an environmental programme and an environmental management system. It is to be developed on the basis of the initial environmental review, and is defined as "a co-ordinated set of measures of various kinds aimed at protecting the environment" (Art 2(b)) – one of the less helpful definitions in the Regulation. The definition of "environmental programme" is a little more useful. It means "a description of the company's specific objectives and activities to ensure greater protection of the environment at a given site, including a description of the measures taken to achieve such objectives and the deadlines set for implementation of such measures" (Art 2(g)). One way of rationalising these definitions might be to say that the environmental programme for a site is a sub-set of the overall environmental policy of the company while the environmental protection system for that site is a sub-set of the company's overall environmental management system.

Environmental policy and objectives

"Environmental objectives" are defined as "the detailed goals, in terms of environmental performance, which a company sets itself" – while "environmental policy" means "a statement of a company's overall aims and principles of action with respect to the environment" (Art. 2(f) and (c)). The company's environmental policy, is therefore, seen as its overall environmental philosophy. In companies which adopt "mission statements" one imagines that there will be an environmental "mission statement" which will constitute the company's environmental policy for the purposes of the Regulation. The first step in putting the policy into a more practical and measurable form is the requirement that the policy lead to the adoption of environmental objectives. It is to be noted that these are defined as *detailed* goals which the company sets itself. It is inherent in the overall scheme of the Regulation that those objectives should be expressed in such a way as to enable their achievement, or non-achievement, to be measured. For example, one of the "good management practices" referred to in Annex I is that "the company shall establish and update the procedures and actions to be pursued in the event of detection of non-compliance with its environmental policy, objectives or targets."

Environmental management system

Having defined the company's environmental policy and environmental objectives, the next step is for the company to set in place its "environmental management", that is "the aspects of the overall management function that determine and implement the environmental policy" (Art. 2(d)). The organisation of the management function varies significantly between different companies in the same Member State, and even more significantly between companies in different Member States and a wide variety of environmental management could be expected to result. However, the purpose of the Eco-audit Scheme is to allow a measure of comparison and beneficial competition and this requires some degree of standardisation. The Regulation, therefore, requires the Commission to request the appropriate European standardisation bodies to develop and adopt standards for environmental management systems based on the criteria in Annex I, taking into account the objectives, principles and provisions of the Regulation (Art. 4). The intention is clearly to standardise the type of environmental management system adopted by each company. "Environmental management system" is defined as "The organisational structure, responsibilities, practices, procedures, processes and resources for implementing environmental management" (Art. 2(e)).

The importance of this standardisation for an auditing scheme cannot be over-emphasised. The primary purpose of an audit, as it is generally understood, is to establish that a company is complying with certain accepted standards – for example, generally accepted accounting principles and practices. If an audit is to form a useful basis of comparison between companies, the definition of the standards which the auditor is seeking to apply cannot be left to individual companies, otherwise there is no element of objectivity in the comparison. While a significant element of the environmental audit will always be the extent to which the company complies with its own objectives, the moves towards a standardisation of environmental management systems will almost inevitably entail a move towards a standardisation of objectives.

Site specific policy and implementation

Having adopted a general policy and general objectives and put in place a management system the company must now apply the policies and systems. The Regulation envisages application of the policy and system by reference to "sites" (as is discussed below). As the Eco-audit scheme itself relates only to the evaluation and improvement of the environmental performance of *industrial activities* the sites in question are industrial sites. Site is in fact defined in purely industrial terms. It means "all land on which the industrial activities under the control of a company at a given location are carried out, including any connected or associated storage of raw material, by-products, intermediate products, end-products and waste material, and any equipment and infrastructure involved in the activities, whether or not fixed". The extent to which the Regulation is site-orientated could not be emphasised more clearly than in the definition of "company" which means "any organisation which has direct operational control over activities at a given site" (Art. 2 (l)). Industrial activity is itself defined as "any activity listed under Section C and D of the classification of economic activities in the European Community (NACE rev.1), as established by Council Regulation (EEC) No. 3037/90, with the addition of electricity, gas, steam and hot water production, and the recycling, treatment, destruction or disposal of solid or liquid waste". The scheme is, therefore, targeted initially at companies in the manufacturing sector. It is also clear, therefore, that the definition of a company does not extend solely to the entities which are normally described under that heading in, for example, English law. If a partnership controlling a site, or possibly a joint venture between two companies, could be described as a separate organisation then it would fall within the terms of the Regulation.

Article 3 explains the participation in the Eco-audit scheme, the commitments by companies participating and the participation procedure. Paragraph 2 of Article 3 states:

> "Companies which decide to participate in the Eco-audit scheme for a given site shall, after having carried out an initial environmental review of the site, established an environmental protection system, and prepared an environmental statement, submit the first environmental statement validated by an environmental auditor, including the information referred to in Annex IV, to the competent body referred to in Article 10."

Whereas, therefore, the definition of environmental policy and environmental objectives is at a general company-wide level, the Eco-audit scheme is primarily concerned not with a company's general policy or objectives but with the policy and objectives which it applies to specific named sites. One of the most important things to remember about the use of the Eco-audit logo which is mentioned below is that it must be accompanied by one of the following phrases (which are set out in Annex III) as appropriate:

> "All our production sites in the EEC participate in the Eco-audit scheme.
> All our production sites in [name of Community Member State] participate in the Eco-audit scheme.
> The following production sites of our company participate in the Eco-audit scheme."

All these statements refer to compliance by specific sites not by the company as a whole. Since the scheme is site specific the same steps must be repeated for each site the company wishes to have approved. The mechanism for applying the company's environmental policy to a site is set out in some detail in paragraph 1 of Article 3. They are as follows:

- "to establish and implement, in the light of the results of an initial environmental review, in line with the criteria in Annex I and ensuring appropriate participation by workers and their representatives, an environmental protection system applicable to all activities at the site,
- to carry out, or have carried out, in compliance with the criteria in Annex I, environmental audits at the sites concerned,
- to examine, at the highest management level, all the findings of the audit and, in the light thereof and within the framework of their environmental programme, set objectives and measures aimed at continuous improvement of their environmental performance,

- to prepare an environmental statement specific to each site audited with the same frequency as the environmental audit,
- to have the environmental statement validated by accredited environmental auditors to confirm that the data and information contained therein are reliable and that all the significant environmental issues of relevance to the site are covered in sufficient detail,
- to submit the environmental statements validated by accredited environmental auditors to the competent body and keep them at the disposal of the public."

In applying the Eco-audit scheme to a particular site the company, therefore, has to go through the following stages:

(i) first, carry out an additional environmental review of the site;
(ii) establish an environmental protection system applicable to the site;
(iii) prepare an environmental statement specific to the site;
(iv) have the environmental statement validated by accredited auditors; and
(v) submit the environmental statement and the auditors' validation to the competent body and keep them at the disposal of the public.

The initial environmental review is defined as "an initial comprehensive analysis of the environmental issues, impact and performance related to activities at a site including in particular the aspects referred to in Annex I." The issues referred would be those in section B of Annex 1 which are:

- "assessment, control and prevention of the impact of the activity concerned on the various sectors of the environment,
- energy management, savings and choice,
- raw materials management, savings, choice and transportation, water management and savings,
- waste avoidance, recycling, re-use, transportation and disposal,
- selection of production processes,
- product planning (design, packaging, transportation, use and disposal),
- prevention and limitation of accidents,
- staff information, training and participation on environmental issues,
- external information and public participation, including dealing with public complaints."

Environmental auditing and the environmental statement

The environmental protection system incorporates the environmental audit, making it but one part of a comprehensive system. As such the audit serves as a management tool having two goals:

- to improve a company's environmental performance; and
- to provide a credible basis for that company in publicising information about its environmental performance.

As a point of reference to assess a company's performance the audit is to determine whether:

- the best available pollution abatement technologies are used;
- the EC and national environmental standards are met; and
- "good management practices" are followed.

The culmination of the environmental audit is the preparation of the environmental statement. The environmental statement is dealt with in considerable detail in Article 5 which requires that it must be "Written specifically for the public in a concise, non-technical form."

Paragraph 2 of Article 5 sets out what the environmental statement shall include:

> "– a description of the company's activities at the site considered,
> – a detailed assessment of all the significant environmental issues of relevance to the activities concerned,
> – a summary of the figures on pollutant emissions, waste generation, raw material, energy and water consumption and other significant environmental aspects, as appropriate,
> – a presentation of the company's environmental policy, programme and specific objectives for the site considered,
> – an evaluation of the environmental performance of the protection system implemented at the site considered,
> – the deadline set for submission of the next statement."

Presumably in the case of the first statement the penultimate requirement, that is, an evaluation of the performance of the environmental protection system will be an indication of the intention rather than a true measure of performance.

This statement ultimately becomes a public document. Its contents, therefore, are of vital importance for the company, the governmental authorities and the public.

Having prepared its environmental statement and had it validated by environmental auditors the company files a copy of the statement with the competent body designated by the relevant Member State. That body then includes the site on a list of sites participating in the Eco-audit scheme. That list is updated regularly and is published in the Official Journal of the European Community every year.

To promote public confidence in a participating company's environmental performance the novel requirement of a validation step has been added. In the United Kingdom the DTI has stated that:

> "the validated statement will be judged on the accuracy of its contents and not on the quality of environmental performance reported. The effectiveness of this approach relies on public pressure to question poor performance or demand more information. Companies will be de-registered if they fail to submit further environmental statements as previously agreed."

Accreditation of environmental auditors

Article 7 deals with the accreditation and monitoring of environmental auditors, in respect of which the Regulation mandates each Member State to develop a system for accreditation.

Paragraph 1 states that:

> "Member States shall ensure, within 12 months of the entry into force of this Regulation, that accreditation systems responsible for laying down and implementing appropriate conditions and procedures for the accreditation of environmental auditors and for supervising their activities are fully operational. To this end, Member States may either use existing certification institutions or designate or set-up other appropriate bodies."

Who should be accredited as an environmental auditor has clearly proved to be a sensitive matter. There has been considerable concern that, as has happened in the United States, inexperienced environmental consultants would emerge, holding themselves out to have much greater expertise than indeed they really have. An initiative in the United Kingdom to overcome this problem has been seen in the proposals for an accreditation scheme for environmental auditors, which have been issued by the Institute for Environmental Assessment (IEA).

IEA was established some two years ago in a first effort to achieve some quality control within the environmental consulting profession. Initially the intention was to improve the quality of environmental impact assessments. Simultaneously, the Association of Environmental Consultancies (AEC) was brought into being and launched in April 1991. AEC's objective was to assure the quality of environmental assessments and audits by ensuring that its members had adequate quality control systems. Subsequently IEA expanded its role to encompass auditing, the AEC issued a code of practice on environmental auditing and the British Standards Institution (BSI) prepared a sector guidance note for environmental consultancies under the British Standard on quality management systems. The IEA initiative, supported by AEC and BSI, was a registration scheme for environmental auditors, *i.e.* individual consultants and in-house auditors within an industry.

The scheme will be administered by a National Environmental Auditors Registration Board. Registration will be in one of three categories based on an individual's academic and professional qualifications, training and practical experience. To be a professional Environmental Auditor 10 points are required, 30 for an Environmental Auditor and 50 for a Principal Environmental Auditor. 50 points are equivalent to at least 180 days experience. The point system is such that academic qualifications may reach a maximum of 5 points, 5 for membership of professional institutions, 5 for training experience and 1 point for each five days of relevant environmental auditing experience.

Under the initiative all candidates would have to sign a professional code of practice, committing them not to undertake work beyond their professional capability. Principal Auditors would have to submit a 5,000 word dissertation, subject to an oral examination, on an environmental auditing topic. According to IEA's Director, Dr Tim Coles, this should provide a quality check which ensures that individuals have achieved the necessary number of days' experience and also ensures that sub-standard auditors do not qualify.

There will be a register of accredited auditors which will be available to the public. Registration would be subject to annual renewal, which would entail continued demonstration of professional development. It is intended that the EC Regulation on Eco-auditing will promote the use of accredited audit verifiers. The IEA has proposed that the register of accredited auditors could be used by companies to locate suitable verifiers, as well as by insurers who need reliable audits of companies that are seeking to obtain environmental impairment liability cover (a subject that is dealt with in Chapter 8).

This initiative should be borne in mind, particularly when considering paragraph 2 of Article 7, which states:

> "The competent bodies, the industrial confederations concerned, non-governmental environmental and consumer organizations and trade unions shall be entitled to participate in setting up and directing, in line with the rules and procedures established by the Member States, the accreditation systems referred to in paragraph 1 . . .."

It is also relevant to the fact that, under paragraph 4, auditors accredited in one Member State may perform validation activities in any other Member State. It will no doubt prove difficult for the system of accreditation to be developed throughout the Community within the 12- month period, especially if each Member State's accreditation system must first be approved by the Commission. So a situation may well develop in which the Eco-audit is available initially in some Member States, but not others.

Article 8 deals with the list of accredited environmental auditors, while Article 9 at paragraph 1, provides that:

> "The bodies designated for this purpose by the Member States shall establish a list of sites participating in the Eco-audit scheme and shall update this list regularly."

Lists of accredited environmental auditors are also to be published in the Official Journal of the European Communities.

The logo

It is intended that participants in the scheme might be encouraged by the opportunity to use an Eco-audit logo, sometimes described as the "Cog and daisy", which is referred to in Article 11 for use by qualifying sites. Paragraph 1 of Article 11 provides:

> "The logo . . . may be used in relation to sites participating in the Eco-audit scheme. The logo shall always be accompanied by a statement of which sites are participating.
>
> The logo may be used:
>
> – on the company's environmental statements,
> – on the company's brochures, reports and information documentation,
> – on the company's headed paper,
> – for the company's advertisements, on condition that such advertisements contain no reference to specific products or services.

The logo may not be used on products, product packages or for product advertisement.
The logo shall always be accompanied by one of the phrases in Annex III as appropriate."

Eco-audit logo

The logo can be used on the company's environmental statements, brochures and reports, headed paper and advertisements. It is important to note in the context of advertisements, however, that the advertisements must contain no reference to specific products or services. The Commission is obviously concerned to avoid advertisements suggesting, by implication, that specific products or services of the company have been approved or audited by the Commission. The significant restriction prohibiting the logo's use on a company's products, product packages or producer advertisements should be emphasised bearing in mind the eco-label initiative (discussed below). There has been some concern that confusion may arise over the two logos, which, presumably, only experience will resolve.

If the company fails to submit a validated environmental statement within the deadline which it announced in its last statement then the relevant site will be deleted from the list of sites maintained by the regulatory authority and the Commission. Furthermore, if any particular site fails to comply with the requirements of Community and national environmental regulations, the company can be prevented from using the Eco-audit logo in respect of that site (Art. 11, para 2). Were it not for this last regulation, a company would be entitled to use the Eco-audit logo even though the auditors' report consistently recorded that the site failed to comply with applicable environmental laws. Currently the proposal intends that audits should take place every one to three years depending upon the circumstances in Annex I.

Assistance to smaller and medium-sized companies

The promotion of the Eco-audit scheme is dealt with in Article 12 of the Regulation. It is self-evident that small and medium-sized companies require some additional incentives, although concern has been voiced that this may result in dual standards. Under paragraph 1 of Article 12:

> "Member States may take action to promote the Eco-audit Scheme, provided such action is in compliance with Community law, in particular, by:
> (a) in the case of small and medium-sized enterprises, establishing or promoting assistance measures and, possibly, structures aimed at providing such firms with the expertise and support needed in order to comply with the rules, conditions and procedures defined by this Regulation and, in particular, to set up environmental protection systems, conduct audits and prepare and validate statements;
> (b) simplifying and/or reducing the practical verification and inspection formalities under certain conditions, in particular for small and medium-sized enterprises;
> (c) advertising and informing the public of the Eco-Audit scheme.
> Member States shall inform the Commission immediately of the measures they take within the framework of this Article."

How the rather imprecise language of sub-paragraph (b) of paragraph 1 will be applied by individual Member States is unclear, but certainly the provision is significant. Paragraph 2 of the Article continues:

> "The Commission, having consulted the two sides of industry, shall present proposals to the Council aiming at greater participation in the Eco-audit scheme by small and medium-sized enterprises, in particular by providing information, training and technical assistance."

Possible review

Article 13 of the Regulation provides for the revision of its terms four years after its entry into force. Since the Commission is committed to the success of the scheme and the review will be "in the light of the experience gained during its operation", there may well be more of a move towards a compulsory system. In any event, in view of the commercial pressure and trends towards "being seen to be green", it would seem that, regardless of whether or not the scheme is reviewed at a later stage and determined to be made mandatory, those companies that become involved in registration and participation in the Scheme could find themselves in a competitively advantageous position. No doubt the practical experience gained through the operation should highlight the scheme's weaknesses and clarify its more confusing provisions. What remains controversial is the questionable effectiveness of a voluntary programme, and its inter-relationship with other EC legislation.

Future considerations

The value of an environmental audit as a management tool is well documented. Until very recently, however, little has been known about the actual practices of industry regarding the use of such a tool.

A survey of approximately 500 companies doing business in England that has recently been completed has revealed some useful information on the actual use of environmental audits. The survey has found that 45% of all respondent companies currently conduct an environmental audit and a further 27% intend to do so within the near future. In view of the fact that there is no existing domestic or EC legislation or regulatory requirement, this decision is an entirely voluntary one, with the survey showing a relatively high level of voluntary use.

Whereas, however, the trends indicated by the survey are in favour of auditing, it should be remembered that participation in the Eco-audit scheme is very complex and expensive. For instance, the scheme requires the development of a comprehensive environmental protection system as well as the hiring of an independent auditor, and, as seen, any deficiencies discovered through the audit should be corrected for the reasons shown above. Even though the Commission has made special provision to encourage participation by smaller companies, the costs are still likely to deter many companies and particularly the smaller ones that lack the necessary technical sophistication and budgetary capacity. Any assistance, therefore, will need to be significant: this assistance will also require some level of uniformity throughout the EC. As it is, the Union of Industrial and Employers' Confederation of Europe (UNICE) has expressed concern "that the proposal in Article 12(1) to lighten and simplify control and inspection for SME's alone represents dual standards." UNICE considers that the Regulation should be generally applicable and that the burden of the auditing activity should be such that individual enterprises can cope with it, and that therefore, such a clause should not be necessary.

Another unresolved issue remains the debate over the approach to environmental audits. Companies in general regard environmental audits as internal management tools. Few even share the audit findings with their shareholders, much less the general public. Getting over the hurdle of the matter of public disclosure will not prove easy. A useful comparison can be seen in the US and industry's response to the Valdez Principles. The Tenth Principle requires companies to "conduct and make public an annual self-evaluation" over their progress and implementing of the Principles and to work towards annual environmental audits which the company agrees to "make available to the public". It is considered that this Principle has been the main deterrent for companies to sign up: by the end of 1990 only 17 companies, of 3,000 approached, had become signatories.

Another issue to be watched relates to the interaction between the Eco-audit and other initiatives, in particular the proposed Civil Liability Directive.

Although the interaction with this Directive still remains hypothetical – and indeed only affects the area of waste management – it could still involve difficult financial repercussions in the event of the Directive's adoption. Having said this, those companies which are presently involved in internal audits should already have some idea of the costs that would be involved in rectifying their environmental problems. They should also be aware of the savings that flow from improved environmental performance. Companies that find this cost savings ratio favourable may find the scheme attractive. What a company may do is to try to conduct the Eco-audit in two stages, the first stage being an internal preliminary assessment of the need to improve its environmental performance; the second being carried out only if the cost/benefit analysis shows in favour of going ahead. It should be mentioned that UNICE has also raised its concern that the Eco-audit does not comply with the usual concept of environmental auditing as an internal management tool. It has taken on board other initiatives, such as in the context of the International Standards Organisation (ISO), and the discussions for various environmental management standards to support the view that an environmental management system should ensure industry's environmental policy and programme are clearly understood, without the need for an environmental statement being based on the environmental audit.

Nevertheless, it may well be that legislation at the Member State level may require disclosure of environmental information comparable to that required under the Eco-audit scheme. Likewise, national legislation may impose environmental requirements which make environmental audits hard to avoid. It has to be said that this type of legislation could result in more companies conducting "voluntary" audits.

Another factor relates to the psychological security that management can achieve through having full knowledge of its company's environmental performance, particularly now that staff and customers are more environmentally aware. Although this may well be unquantifiable, this could lead to more audits than a straight cost/benefit analysis would indicate.

The value of the logo, which has been referred to above, is particularly relevant. In this respect, it should be noted that the Eco-audit scheme does not operate within an entirely "voluntary model": rather, it seeks to create positive market forces as economic incentives for participation in the scheme. The logo represents an indication of a company's environmental soundness, and public education regarding the scheme means that market-based incentives are evident. In addition, there is the added incentive in the form of the vaguely worded provision which indicates that participants may have greater freedom from regulatory control, as well as the requirement that Member States should come up with a programme to aid small- and medium-sized companies. To the extent that the latter programme involves subsidies or taxing advantages, there is potential for a direct and immediate economic incentive. Altogether, these incentives, as well as the force of the confidence inspired by going through the Eco-audit and from the point of view of shareholders, consumers and the general public, mean that companies may feel that they must conduct audits, particularly those in certain industries or regions.

It remains to be seen, therefore, how far the Eco-audit scheme will prove successful and to what extent it will fit into the European Community's combination of voluntary regulatory and market-based approaches. No doubt, further modifications to the proposal will occur following the period of consultation and consideration by the various working groups and COREPER. Hopefully, such modifications will enable the optimum participation in the Eco-audit without eroding its voluntary, market-based philosophy. The document will of course emerge in its final form following the Parliament and Council sessions and should be noted accordingly.

Last but not least, there is no doubt that a successful Eco-audit scheme would not only improve the environmental awareness and performance of companies, but also involve a major contribution to the well being of our environment.

CONTENTS OF CHAPTER 3

Corporate Policy and Environmental Management

Chapter 3
Corporate Policy and Environmental Management

At the corporate level the fundamental elements of environmental awareness are an environmental policy and an environmental management system. Many large and medium-sized companies have already developed such a policy and system. In the next few years all companies of any size will develop them in some form. Developing systems appropriate to different sizes of companies will involve a difficult balance between environmental protection and commercial bio-diversity.

The immediate subject of an environmental policy will be the company's compliance with existing and proposed environmental regulations. However, an effective (and attractive) policy will involve some element of forecast and will be attuned to social and consumer pressures which are generally in advance of such regulations.

An environmental policy is ineffective – and may expose the company to additional liability – unless it is accompanied by an effective environmental management system. Such a system must facilitate the measurement of performance both by internal management and by external auditors. The adoption of an effective environmental management system requires co-operation at all levels of the company's workforce.

The Eco-audit Regulation

The Eco-audit Regulation, as has been discussed in Chapter 2 (and subject to comment in Epilogue), envisages that every company which wishes to join the Eco-audit scheme will have an environmental policy and an environmental management system applicable to the company as a whole. The Regulation is strictly concerned only with the environmental management of industrial sites by industrial companies. "Company" is defined to mean "any organisation which has direct operational control over activities at a given site" (Art 2(k)) so it is not confined to "companies" within the meaning of the UK Companies Act, although this chapter discusses the application of the Regulation and other environmental obligations with companies in mind. The Regulation makes it clear that the development of environmental protection systems for sites will normally proceed within the context of a broader policy for the company as a whole.

> "The corresponding documents will explain how the environmental policy and the environmental management system for the site relate to the policy and systems of the company as a whole and will include a statement of the environmental policy conducted by the company" (Third paragraph of section A of Annex 1).

It would be possible in theory for a company to comply with this obligation by merely stating that there was no company environmental policy or system, but it seems to be of the essence of environmental policies adopted in practice that they are imposed "from the top down". Further, the preamble suggests that while industrial companies are in the vanguard of developing environmental policies at the moment, sooner or later there will be similar regulations which will apply to non-industrial companies:

> "Whereas companies, in particular industrial companies, are faced with constantly growing, far-reaching and complex tasks and responsibilities related to the assessment, control, prevention or reduction of the environmental impact of their activities;
>
> Whereas these tasks and responsibilities call for companies to establish effective environmental protection systems;
>
> Whereas the establishment of environmental protection systems and the use of environmental auditing in this context are more advanced in the industrial sector;"
>
> (Tenth, eleventh and fourteenth paragraphs of the Preamble to the Regulation).

Even if the Regulation is not extended, the growth of environmental responsibilities imposed upon companies by national and international legislation will make such policies and systems necessary. Directors will increasingly be called to account on the subject of their environmental management, both by regulators and by shareholders. The ability to demonstrate an effective environmental management system will be an important, if not an essential, element in showing that they have discharged their duties.

The present position

In 1989 *Business International* conducted a survey of Chief Executive Officers of over 100 leading companies operating in Europe. Some of the companies were based in North America, others in Europe. Overall, 48% of the companies had already published an environmental policy. However, the figure for those of the surveyed companies which were based in North America was 100%. It is a common saying that trends which arise in North America, and particularly the United States, will be followed in Europe five years later. In the environmental field as a whole that statement is a gross over-simplification of a complex relationship, but as regards corporate policy it is almost certainly correct.

There is little reason to doubt that since 1989 the proportion of larger European-origin companies which have published an environmental policy has increased and that it will, in due course, reach 100%.

An evolutionary process - the example of ICI

Publication of the company's first environmental policy is only the beginning of the process. Leaving aside for a moment the questions of implementation and audit, the experience of forerunners in the market indicates that a company's environmental policy will undergo a more-or-less continuous process of development. It will be subjected to various pressures. On the one hand, implementation may reveal that a particular objective of the policy (or the time-scale for its realisation) was too ambitious; on the other, new regulations may require a planned implementation programme to be accelerated. The phasing out of CFCs is an example of ever-shortening timescales being imposed by regulators in response to the growth of scientific evidence on the damage caused to the ozone layer.

Most of all, the increase in knowledge and broadened perspective which results from a detailed examination of the interaction between the company's activities and the environment may lead a company which initially adopted an environmental programme as a matter of self-defence – to monitor compliance with burgeoning environmental regulations – to become an evangelist of sustainable development. The distinguishing factor between companies will not be whether or not they have an environmental policy, but the depth of their environmental commitment.

The fossil evidence of the evolution of a corporate environmental policy can be studied in the Annual Reports of Imperial Chemicals Industries plc (ICI). As a major chemical company ICI's impact on the environment is both more direct – and more visible – than that of most other companies. It is also one of the United Kingdom's leading companies and maintains a high public profile. That combination of factors means that it has long been a favourite target of environmental campaigners in the United Kingdom and that its relationship with the environment is subject to considerable public scrutiny. It is not surprising that it is in the vanguard of UK companies adopting visible environmental programmes. How has it presented its environmental programme to its shareholders and the public?

In 1987, ahead of most UK companies, ICI created a "Group overview responsibility at Executive Director level" for Safety, Health and Environment. Although the policy is not stated in the 1987 Annual Report (which devotes half a page to "Safety, Health and Environment") ICI adopted its first environmental policy at that time:

> "It is ICI's policy to manage its activities so as to give benefit to society; this entails ensuring they are acceptable to the community and that any adverse effects on the environment are reduced to a practicable minimum."

This is an interesting statement. The tone is muted. There is no suggestion that adverse effects on the environment are to be eliminated; the commitment to reduce them to a practicable minimum anticipates that some adverse effects will always remain. The reference to practicability suggests that financial and other factors may take precedence over such reduction.

On the other hand, the apparent commitment is astonishingly wide. Does ICI really manage its activities so as to give benefit to society? What if the interests of society and its shareholders are in conflict? What does "society" mean in this context anyway? Can ICI really ensure that all its activities are acceptable to the community? What does "community" mean in this context? Does it mean the local Chamber of Commerce, the local ICI employees and their families, the occupants of the estate next to the chemical plant, or the local branch of Friends of the Earth?

The 1988 Report, entitled "Making the World a Better Place", sets out the policy in similar language:

> "ICI's policy on the environment is to manage its activities so that they are acceptable to the community. We co-operate fully with the authorities to ensure we meet out legal obligations; and we adopt best practice to keep adverse effects on the environment to a practical minimum. This policy inevitably incurs substantial costs, and increasing environmental pressures are likely to narrow the business opportunities in some areas . . .. We remain committed to developing ways of assessing our safety and environmental performance. Under a reporting-back system, territories and businesses report to the Executive Directors on their targets, action plans and performance, alongside financial plans and forecasts."

The environment together with safety and health, still merits half a page.

In the 1989 Report, the theme is subject to significant variations:

> "ICI's policy is constantly to improve its environmental performance. This involves managing its activities so that they are acceptable to the community, adopting best practice to keep adverse effects on the environment to a practical minimum, and co-operating fully with relevant authorities to ensure that legal obligations are met. The Group's reporting mechanisms enable it to track environmental performance and each site has developed improvement plans. Progress against these plans is regularly monitored . . . ICI believes that its financial position will not be materially affected by the fulfilment of its environmental responsibilities."

There is a new tone to this statement. Firstly, it begins with an oblique admission that ICI's environmental record is not perfect: "ICI's policy is constantly to improve its environmental performance." Secondly it acknowledges the cost of taking environmental performance seriously: not just a narrowing of business opportunities but a potential – if disclaimed – threat to the Group's financial position and competitiveness. The space occupied by the section has increased to one page.

In the 1990 Annual Report, a feature entitled "Towards A Better World" appears immediately after the Chairman's Statement in the most prominent position in the report. The feature begins by restating the ICI environmental policy with subtle (and not-so-subtle) changes:

> "ICI seeks to ensure that all its activities are acceptable to the communities of which it is part and that the environmental impact of its operations is kept to a practicable minimum."

The commitment is considerably less forthright than that contained in the 1987 statement of policy. The reference to activities being of benefit to society has gone. "Seeking to ensure" that activities are acceptable is a lesser obligation than "ensuring" they are acceptable. A reference to the "environmental impact" of operations is rather less graphic than a reference to their "adverse effects on the environment" but means much the same. Interestingly, this change of phrase leads to the later policy *appearing* to be more positive than the earlier policy although an analysis of the wording shows the contrary. The "practicable minimum" is still there as a substantial qualification. Most subtle of all, the change from a reference to "the community" to a reference to "communities of which [ICI] is part" can perhaps be seen as a suggestion that ICI will apply different standards in different communities. Later in this chapter we will consider the dangers of being too specific in stating an environmental policy. ICI appears aware of those dangers.

Does this mean that ICI's environmental commitment has diminished over the three years? ICI has its own explanation of what has happened:

> "Two years ago the ICI Annual Report announced that the Group's businesses were required to report regularly on how they had improved their environmental performance. This formalization of the reporting process has made it possible in the intervening years to build up an increasingly accurate picture of ICI's environmental record around the world. With this in place, ICI has been able to set specific objectives for improvement across the Group.
>
> The objectives in question go beyond the requirements of the law – compliance being the minimum standard ICI sets itself. The intention is that ICI's environmental performance should become one of the best in the industry. There are four objectives in all.
>
> Firstly, all new ICI plants will be built to standards capable of meeting all regulations that might reasonably be expected in the most environmentally demanding country in which ICI will operate that process.
>
> Secondly, ICI will reduce waste from its operations by 50 per cent by 1995, paying special attention to that which is hazardous.
>
> The third objective is to establish an even more rigorous programme for conserving energy and resources, paying particular heed to actions that will safeguard the environment.
>
> The fourth objective is to set up waste recycling programmes, not only in-house but also in collaboration with customers.
>
> The announcement of these new objectives in November 1990 marked not so much a change in policy as a gear change in ICI's environmental effort. What is new is the setting of Group-wide targets against which performance can be measured – which in turn makes possible realistic plans for meeting them.
>
> ICI's capacity to solve environmental problems – and the seriousness with which it takes its environmental responsibilities – make the new objectives entirely realistic. Nor are these objectives fixed. Rather they are the next set of targets the Group wishes to meet in its continuing programme of improvement. When these are achieved, or when new technology makes tougher objectives possible, the list will change."

The process of environmental review carried out between 1988 and 1990 has had two effects. First, the sweeping commitments in the 1987 statement of policy have been silently abandoned. It is questionable whether the changes can be described as only a "gear change" but it is unlikely that the abandoned objectives would ever have been "practicable" in any event: what the 1990 statement *says* is what the 1987 statement probably *meant* in practice. Secondly, as the company says: "Group-wide targets [have been set] against which performance can be measured – which in turn makes possible realistic plans for meeting them."

Both changes can be seen as a result of the same process – the testing of inchoate environmental aims against the practical realities revealed by an environmental review – and there seems little doubt that, overall, all interested parties are better served by the 1990 policy (and the management system that gave rise to it) than by the 1987 policy. The substitution of a realistic policy and well-defined objectives means that ICI's management has a means of measuring and reviewing the company's environmental progress; but it also means that regulators and environmental campaigners have more information about the company's activities and a clearer framework in which to conduct a constructive dialogue. It is easier to persuade a company which is committed to a 50% waste reduction to move to 90% waste reduction than it is to persuade a company that a commitment to activities "acceptable to the community" entails an obligation to cut waste reduction by 90%.

By the 1991 Annual Report, the section devoted to Safety, Health and the Environment in the main report has been reduced to two pages but those two pages merely summarise an additional report sent to all shareholders with the statutory part of the Annual Report. That additional report is effectively an audit report on ICI's achievement of its four specific objectives and a statement of revised environmental goals for the future. The process of refining policy in the light of review had advanced further:

> "A high priority in 1991 was to refine the Group's ability to monitor its environmental performance. Systems were developed for measuring environmental emissions more precisely. At the same time, greater effort is being made to inform the public of what is being emitted from ICI sites."

Looking only at the physical prominence given to the environment in ICI's annual reports, it has become more salient and more specific and consequently more measurable. Allowing for the fact that companies in their annual reports, like politicians in Parliament or Judges in the House of Lords, seek to persuade us that their present position, however much it may appear a radical departure from the past, does not really represent any change of policy, it is difficult not to conclude that ICI in 1992 views the relationship of its activities and the environment in a manner which is qualitively different from the manner in which it viewed that relationship before 1987.

Various reasons for this change can be suggested. In November 1992, ICI announced that it was withdrawing from the production of chlorine at Hillhouse site, near Blackpool, at least partly because of the anticipated cost of environmental compliance in future years. ICI, being an international group consisting of some nine businesses with approximately 400 sites world-wide, has subsidiaries which must comply with the environmental regime of non-EC countries. The influence of stricter regimes in those countries is significant, both because of the obligations it imposes upon the relevant subsidiaries and because, in giving an indication of how EC legislation might develop, it allows for anticipatory action. The aim is to integrate quality management in health, safety and environmental matters through a group wide programme world-wide, to ensure a consistently high standard of environmental practice. There are now 19 internally published standards (with 150 guide-lines) of which the final one is "thou shalt audit". Auditing is seen as comparing what is with what should be; the purpose is not policing but instead to provide positive assistance to management control by identifying areas for improvements, thereby assuring compliance. (See also Chapter 2.)

Voluntary compliance with foreign countries' stricter standards can also be a selling point in itself: "The process [for producing a particular chemical] meets all environmental standards anywhere in the world" appears in the 1991 Annual Report. There can be positive spin-off in confidence among shareholders, raising of awareness and improving competitiveness.

A very interesting example of companies competing to comply with standards which are of marginal application to the main market for their products lies in the field of computer monitors. Over the past one or two years, several EC, Japanese and American companies have launched monitors which are prominently claimed to comply with Swedish standards for electromagnetic emissions. This is not because they anticipate massive sales of such monitors in Sweden but because consumers in other countries believe that Swedish safety standards are particularly stringent and would prefer to buy a monitor complying with the standards imposed by Sweden than they would with the standards imposed by their domestic governments.

However, legislation – even anticipated legislation – cannot explain the size or the speed of the change. Nor can pressure from environmental groups and consumers – although both of these must be important. It is suggested that the process of investigation, and the experience gained as a result, has in itself led to a greater degree of commitment. One of the reasons may be found by returning to the production process which "meets all environmental standards anywhere in the world":

> "Setting out to cut the waste involved by 50 per cent, the business in fact achieved a 90 per cent reduction, saving itself £70,000."

The past philosophy of doing the least that one can get away with has been replaced by the philosophy that if one does only the minimum one will not survive in the long-term.

Caveat

However much ICI may have moved towards greater openness it has not (at least in its published reports) taken what is perhaps the most important step, the step required by the Eco-audit Regulation: an objective external audit.

Until companies are prepared to obtain, and publish the results of, external environmental audits they run the risk of being accused of "massaging" their environmental records for the marketing effect. However, the employment of an external "auditor" is not, of itself, protection against criticism while the company retains control over the scope of the audit and the publication of the report. For example, Andrew Jack in the *Financial Times* claimed that while the text of the Body Shop's "green book" "gushed over the company's achievements in 47 pages", the statement at the end from the independent consultants, or "verifiers" of the information in the report, revealed that they had only examined one of the company's plants, had not conducted a comprehensive environmental audit, and had not validated figures quoted by the company on raw material use, energy consumption and waste. Most recently ICI has moved in the right direction by embarking upon a company auditor training programme.

Adoption of an environmental policy

What does a company have to do in order to adopt an environmental policy? There are three layers of long-term policy within a company. If the company is looked upon as a building, the three layers represent its foundations, its superstructure and the use to which it is put.

The basic layer – the foundation layer – is found in the memorandum of association of the company. The memorandum sets out the company's "objects": the purposes for which it has been formed. The objects are broadly stated and very rarely change.

The second layer – the superstructure – consists of policies adopted by the company's shareholders in general meeting. These policies are primarily concerned with structural matters – for example, rights attached to shares, the company name, share options schemes – and, as such, are subject to occasional, but infrequent, alteration and "improvement".

The first two layers provide the shell in which the company carries on its business. What most people recognise as a company's "policies", however – in other words, its "business" policies – fall within the third layer. Within this third layer, which is the primary preserve of the directors rather than the shareholders, are many different types of policy – long-term, medium-term and short-term. Examples of long-term policies might be the company's health and safety policy or employment policy. Both of these are likely to be subject to change annually, if not more frequently, but that change will normally be progressive rather than revolutionary. An example of a medium- or short-term policy is a company's car policy. Some of these policies may have been written down; others may not. Some may be public; others may be a matter of internal management.

A company's environmental policy will clearly normally fall within the third policy layer and will be considered further below. As environmental policy is not concerned with the company's structure the second layer of policy can be ignored here, but before looking at the third layer it is worth sounding one cautionary note about the first layer of policy: the company's objects. This requires a small historical diversion.

A company's objects

Since their earliest days, all companies have been formed for a particular purpose – for example, to manufacture chemicals – and have been obliged to state that purpose in some public document, known as the memorandum of association. The stated purposes of a company – there could be more than one – were its "objects". Until the Companies Act 1989 and the cases which led up to it, a company had the capacity legally to do anything which it stated to be one of its objects (provided that object was not in itself illegal) but did not have the capacity to do anything which could not, expressly or impliedly, be brought within those objects. Any act which was beyond its capacity was void and, therefore, a person dealing with a company had to take care that the transaction which was in contemplation fell within the objects of the company. Recent examples of the dramatic effects of this rule can be found in the cases relating to Local Authorities' transactions in "swaps" and other financial instruments. Because the transactions were beyond the authorities' powers, they were not liable to pay the resulting debts (totalling millions of pounds) to the banks which had sold them the instruments.

This rule has gradually been diluted and it is reasonable to say that a third party acting in good faith now need not concern himself with the objects of a company with which he enters into business. The objects clause has not been abolished, however, and it still has significance for the company's shareholders and directors. If the company's directors attempt to do something which is beyond the capacity of the company, as set out in its objects clause, the shareholders can prevent them doing so. If they go ahead regardless, they can be sued by the company, and possibly by its members, for breach of duty.

However broadly the specific objects stated in a company's memorandum may be drafted, the company is likely to have to carry out acts which are not covered by any specific object. This causes little difficulty when the act in question is commercial in nature and clearly ancillary to the company's main business: a capacity to carry out ancillary acts will normally be implied. However, there have been doubts in the past as to the extent to which a commercial company can indulge in charitable or "uncommercial" acts, such doubts being based on the ground that such acts are not clearly demonstrable to be ancillary to the business. To minimise this difficulty, certain common "uncommercial" acts are often included as specific objects of a company: for example, guaranteeing the liability of third parties – particularly other companies in the same group – or granting pensions to former directors' widows; neither of which acts might be considered to be directly for the commercial benefit of the company. The argument is that by making these acts into specific objects of the company, the necessity of showing that they are ancillary to any business purpose is avoided.

Most companies will initially adopt an environmental management policy because it is an effective means of ensuring that the company deals with existing and proposed environmental legislation. Compliance with legislation is, however, only part of the purpose of the Eco-audit scheme and companies which subscribe to the scheme, at least in its early stages, can be expected to be undertaking obligations which are more onerous than those imposed on them by current legislation. To that extent they are undertaking voluntary obligations.

It is not excessively cynical to assume that in most cases the board of directors will consider that the public adoption of an environmentally-aware stance is likely to produce commercial benefits to the company. As the Eco-audit Regulation itself says: "... the public is showing growing interest in information about the environmental performance of companies". However, it is not impossible to conceive of a situation in which the board and the shareholders take different views about the subject. A radical board might consider it to be in the long-term interest of the company to enter into an environmental clean-up programme which far exceeded the requirements of current legislation, on the grounds that by doing so the company would gain an enhanced environmental image with consumers. Current shareholders, on the other hand, might consider that such a policy (which exhausted profits which would otherwise be available for dividend) was speculative and, to that extent, "uncommercial". In such circumstances, it is conceivable that shareholders might seek to block the directors' decision on the ground that the environmental policy was not ancillary to the company's business and was not permitted under the company's memorandum of association.

Directors contemplating the adoption of a radical environmental policy should, therefore, give some consideration to the question of whether a first step should be to seek shareholder approval for the adoption of a new "object": that the company develop and implement an environmental policy, environmental managements schemes etc.

Employee participation

Apart from inserting a new object in the company's memorandum of association, what else does the company have to do to adopt an environmental management policy? One requirement stated in the Eco-audit Regulation which would not necessarily be common practice among UK companies is to ensure "appropriate participation by workers and their representatives" in the formulation of the environmental protection system. In countries with worker directors or works councils there are established procedures for ensuring "appropriate participation by workers and their representatives" (Art. 3(1)). What is appropriate in the UK context is less clear. Employees are referred to in two sections of Annex I. As mentioned above, one of the issues to be covered in an environmental protection system is "staff information, training and participation on environmental issues" and the following appears in the list of "good management practices": "Management, in co-operation with staff representatives, shall foster a sense of responsibility for the environment amongst employees at all levels" (Annex I C(2)). Neither of these references appears to envisage any great employee involvement in formulating policy. Indeed, it is not clear from the wording of the Regulation whether the appropriate participation is intended to be in the formulation of the policy or in the environmental protection system which results from a formulation of the policy. Given that employee participation has provided a constant source of friction between the United Kingdom and the Commission, it is to be hoped that the implementation of the Eco-audit scheme in the United Kingdom will not be delayed by an argument as to the extent of this participation.

In this context it is important to note that the Regulation requires appropriate participation in the case of "environmental protection systems", rather than "environmental policy". It will be recalled that environmental protection systems are site-specific so it might be that the participation referred to is participation in formulation of policy for a particular site rather than for the company as a whole. Nevertheless the trend is toward the greatest possible participation by workers. For further discussion of the Eco-audit provisions and British Standards Institute (BSI) management systems see Chapters 2 and 4.

Review of policy

Once environmental policy and management systems have been adopted for the company as a whole, and environmental protection systems have been adopted for one or more sites, the audit process can begin. The objectives of the audit include, however, "assessing the management systems in place, determining compliance with community, national and local regulations and determining conformity with company policies". The question which will be addressed here is the effect of an audit showing that the company has failed to conform to its own policies or to comply with applicable regulations.

In general no legal consequences follow from a company silently adopting a beneficial policy, but consequences can follow when a company publishes that policy.

Private companies

There is nothing to stop a private company adopting an environmental policy and environmental management system the equal of anything suggested in the Regulation and appointing an accredited environmental auditor to monitor its performance. Suppose then that the auditor reports back to the company's directors, who also represent a majority of the company's shareholders, that they are failing to meet the objectives which they have set themselves including, in some comparatively minor respects, being in breach of current environmental regulations. The directors may well decide that they have been over-ambitious in formulating their policy and that they should modify their objectives so as to concentrate resources on complying with existing legislation and not seek too vigorously to anticipate more onerous future requirements. Alternatively, the company may receive a glowing report from its auditors but found that, due to a decrease in profits, the cost of meeting its ambitious objectives was greater than it can bear in a particular year. The directors could decide to modify the objectives or to postpone their achievement. In either case – assuming that they always comply with the existing law and that there is no difference of opinion between the directors and the shareholders – the directors are free to abandon or modify the company's environmental policies without reference to any third party.

The position would be slightly different if the same company, instead of being run as an enlightened despotism, adopted an element of democracy and involved its employees in the formulation and implementation of its environmental policy. There is then a "third party" who knows the company's environmental policy: the company's employees. The additional constraints imposed by this knowledge are unlikely to be great. The board would be wise always to involve employees' representatives in a decision to change policy.

Disaffected employees who have not been consulted as to a change of policy may be tempted to leak the company's failures to the local press, despite the fact that most employees will be bound by the contracts of employment to observe the company's environmental policy "as decided from time to time" and will, therefore, not legally be entitled to complain at any change. However, if an environmental policy, announced to the company's work force, takes credit for measures increasing the health and safety of the company's employees, consideration should be given to the effect of a future change of policy. For further discussion on health and safety issues see Chapter 6.

Suppose now that the same company, having successfully adopted an environmental policy and secured the co-operation of its workers, decides to go one step further and to join the Eco-audit scheme. Four important categories of people are added to the numbers who are aware of the company's environmental policy. They are the Commission and the regulatory authorities (and behind them the public), the company's shareholders, the company's suppliers and the company's customers.

Regulatory authorities

The effect on the company's membership of the Eco-audit scheme of its failing to issue an environmental statement at the correct time or the revelation of its non-compliance with the Regulation is referred to in Chapter 2. A further relevant point is that the regulatory authorities, who previously might have been unaware of the company's defaults, will receive a written report setting them out in great detail. This has two consequences. The first is that the regulatory authorities will be aware of breaches of which they might not otherwise have been aware. The second is that, because the public and environmental pressure groups will be aware of the availability of information, there will be more pressure on the authorities to take action and their discretion as to whether or not to prosecute might, therefore, be reduced.

Shareholders

It has been assumed earlier that the directors have an identical, or at any rate do not have a difference of, opinion with the shareholders and therefore the effect of telling the shareholders of the policy and any potential changes in it is unlikely to be of any effect.

Suppliers

There have been circumstances in which companies have refused to supply certain purchasers – for example, many companies used to refuse to supply companies in South Africa. It is conceivable that a generation of "green suppliers" could arise who could refuse to supply companies who failed to comply with their published environmental policies. It might well be that in practice, however, the supply contract lost because a company breaches its environmental policy would not have been gained in the first place had the company not published an environmental policy. In other words, if one seeks to gain an advantage by adopting a public environmental policy it is not unreasonable to foresee the potential loss of business if one fails to comply with that environmental policy. In many cases the effect on potential customers will be the main reason for adopting an environmental management policy or joining the Eco-audit scheme. As with suppliers, the risk of losing those customers for failure to comply with the policy is balanced by the fact that at least some of those customers will have been attracted to the company in the first place by the publication of an environmental policy.

Public companies

When one turns to public companies the effect of adopting a corporate policy creates similar problems. In the case of employees, regulatory authorities, suppliers and customers the problems are the same as those referred to in the context of private companies. The major difference, however, is that the affairs of a public company are subject to much greater scrutiny and a public company which is failing to comply with its environmental policy is much more likely to find its goods subject to a public boycott organised by an environmental pressure group than a private company.

Where the risks take on a different nature is in the case of shareholders. In our imaginary private company the shareholders were very much identified with the directors. In most public companies there is no such identification. At the time of writing this chapter it is being alleged that BP expressly or impliedly adopted the public policy of maintaining dividends despite adverse financial circumstances. The company is being sued by shareholders in America who claim that they bought shares on the face of that representation. With a growing interest in ethically aware investment, it is not difficult to imagine that shareholders who buy shares in a company based on its public declaration of an enlightened environmental policy, might commence a similar action against that company if it were shown to be half-hearted in its implementation of the policy, or to vary it in a manner which was considered to be damaging to the environment. This possibility is considered further in the next chapter.

Closing remarks

In summary, the adoption of a public environmental policy might heighten the risk of being prosecuted by regulatory authorities and incur the risk of losing suppliers or customers in the event of a public breach of environmental policy – although in the case of suppliers and customers the risk is probably outweighed by the benefit of adopting the policy in the first place. In particular, it is worth remembering that in the absence of an environmental policy, a company may find it more difficult if not, in the fullness of time, impossible to obtain work. For many years companies have been subject, by law, to a financial audit at the end of each financial year. More recently, Private Members Bills have been laid before the UK Parliament extending compulsory financial audits to environmental and health and safety issues. This has been coupled with more public accountability with a requirement for information such as:

(i) details of a company's health and safety and environmental convictions, including the levels of fines and compensation; and
(ii) details of any personal convictions by Directors (see the Corporate Safety and Environmental Information Bill discussed in Chapter 6).

Further, the ICAEW's research group has recommended that companies should provide a "true and fair" picture of their overall environmental performance annually.

There are also embryonic signs that environmental (and health and safety) policies may become a pre-tender requirement for new work. In the construction industry, for example, a good health and safety record and/or a proper health and safety policy are now occasionally being required as a pre-qualification for tender. Other developments internationally, such as India's new requirement for an annual environmental audit as part of its corporate statutory accounting requirements, emphasise the value of a considered environmental policy. There is no doubt that this is an issue to watch.

CONTENTS OF CHAPTER 4

The Eco-audit Regulation and the Move Towards a World Standard for Environmental Management Systems

Chapter 4

The Eco-audit Regulation and the Move Towards a World Standard for Environmental Management Systems

The interaction with other initiatives

First of all it is useful to recall the Eco-audit scheme's substantive basis. Its "preventive" nature finds ample basis in Treaty law (particularly in view of provisions recently added at the Maastricht Conference). For instance, Articles 130s and 130t of the Treaty of Rome require EC environmental policy to be based on the principles of preventive action, rectification at source, the polluter pays and the integration of environmental considerations into other European policies. The Maastricht Treaty has added that EC environmental protection should be based on the "precautionary principle". In so far as this principle predicts the development of environmental measures which anticipate, prevent and attack the causes of environmental degradation even in the face of scientific uncertainty, this provides a clear endorsement of the Eco-audit initiative and the preventive approach it represents. The scheme similarly finds support in the EC Statement of Environmental Objectives and Principles and in the Fourth Environmental Action Programme which emphasise the need:

- to prevent pollution at source;
- to develop environmentally sound management; and
- to better educate the public through the dissemination of increased information regarding environmental performance.

The Fifth Environmental Action Programme, as we have seen, continues this trend.

The Eco-audit scheme must, however, be seen in the light of other developments both of the European Community and beyond. Those other developments represent not only a background which, as Chapter 1 has shown, set the scene for the scheme itself, but also a world-wide movement towards a more unified approach to dealing with environmental issues and standards, of which the scheme forms just a part. It is, however, worth remembering that as Europe strives for minimum standards, coupled with an encouragement to each Member State to improve those standards (see Art 130T of the Treaty of Rome) there remains the continuing debate as to what is meant by standards. Is it, by way of example, a biological standard (the original approach by the Community based on a human health test), an emission standard (see Air Pollution Directives), a preventive standard (see Environmental Impact Assessment Directive), an environmental quality standard (see Bathing Water and Drinking Water Directives)?

In addition, the Fifth Action Programme encourages a move away from media standards (i.e. specific Directives on water, air and land) to a cross boundary approach to environmental standards (see the proposed Integrated Pollution Prevention and Control Directive), substantially based on the UK Integrated Pollution Control Directive under Part 1 of The Environmental Protection Act 1990) (see also Chapter 7).

In Chapter 1, a number of aspects of this subject have already been touched on. Certain of them are given a further treatment in this chapter. Proceeding from this basis some of the work already under way in pursuit of a more consistent approach at an international level will be examined.

Comparison with the British Standard management system

The most important difference between BS 7750 and the EC Eco-audit scheme is undoubtedly in relation to the scope of the systems. BS 7750 is, as its name states, a "Specification for Environmental Management Systems". Companies wishing to register under BS 7750 must establish and maintain a management system according to the specifications. The Eco-audit scheme proposed by the Commission is much broader in scope than this. Companies wishing to participate in this scheme must establish and maintain an environmental protection system which meets the requirements set out in the Regulation. This includes, but is not limited to, establishment of an environmental management system. Not only is it necessary to introduce a system for *managing* environmental matters, it is also necessary to establish measures aimed at *protecting* the environment. Accordingly, as the British Standards Institution (BSI) itself has stated, although use of BS 7750 may be seen as a first step towards meeting the requirements of the Community Eco-audit scheme, it is not an alternative way of achieving the same result. The difference between an environmental management system and an environmental protection system is a substantial one. Although BS 7750 requires that participants have an environmental policy which includes a commitment to continual improvement of environmental performance, the essence of an environmental management system such as BS 7750 is that there is effective management control over the environmental performance. The standard of performance itself is very much of secondary importance. By way of contrast, the essence of an environmental protection system is that there is a system for improving the environmental performance. Many of the differences which are highlighted below are attributable to this fundamental difference in scope.

A less fundamental but still noteworthy difference in the scope of these systems is that, whilst BS 7750 is intended to apply to all types and sizes of organisation, the Eco-audit scheme is only aimed at companies operating an "industrial activity", as defined in the Regulation (see further the Epilogue).

Establishing the system

The environmental protection system required under the Eco-audit scheme must be established in the light of an internal environmental review. This review must be a comprehensive analysis of the environmental issues, impact and performance related activities at the site in question and must take into account certain aspects stated in the Regulation. *Seemingly, if the environmental review is ineffective in that it does not take into account all of the required issues, the protection system set up in the light of the review will not satisfy the requirements of the Regulation.* Whilst BS 7750 also requires that a similar "preparatory" review should take place before the environmental management system is established, it is clear that BSI does not regard this initial review as part of the specification, because as BSI has stated "it is not an assessable element of an establishment system". Accordingly the adequacy or inadequacy of the preparatory review should not have a bearing upon whether or not certification will follow.

Environmental policy objectives and targets

Both BS 7750 and the Eco-audit scheme require that those participating have an environmental policy that is publicly available. In relation to the contents of such a policy and the setting of objectives and targets, BS 7750 states simply that the policy must include *a commitment to continual improvement of environmental performance* and that objectives and targets shall quantify, where practicable, the commitment to this. BSI intends to produce industry sector guidance notes which will identify the key environmental performance criteria relevant to the particular industry concerned, to assist participants in setting these targets and objectives. The only specific proviso as regards objectives and targets, however, is that they must go beyond mere compliance with all legislative and regulatory requirements. It is important to note that in setting objectives and targets under BS 7750, organisations are specifically allowed to take financial, operational and business requirements into account. Under the Eco-audit scheme, by way of contrast they are expected to ensure that the *environmental impact of all activities is reduced to the minimum.*

Organisation and personnel

BS 7750 contains very specific requirements relating to employees: for example, a company operating BS 7750 will be required, amongst other things to:

(i) define and document the responsibility, authority and interrelations of key personnel;
(ii) appoint a management representative who will have the authority and responsibility for ensuring that the requirements of the Standard are satisfied;
(iii) establish and maintain procedures to ensure that all employees are aware of the importance of compliance with the Standard and their responsibilities in this regard, the potential environmental effects of their work activities and the environmental benefits of improved performance;
(iv) establish and maintain procedures to ensure that employees whose work may have a significant effect on the environment are appropriately trained and that records of training given are kept; and
(v) ensure that work is undertaken by a suitably qualified person.

The Eco-audit Regulation is much less specific in this area but it is potentially more demanding. It provides that organisations must ensure the appropriate participation of employees in the setting up and running of the environmental protection system and must foster a sense of responsibility for the environment amongst its employees. It also requires that the environmental protection system must include the provision of information and training to employees about environmental issues. In countries with worker-directors or works councils there are established procedures for ensuring appropriate participation of employees. What is appropriate in the UK context is less clear. Given that worker participation has provided a constant source of friction between the United Kingdom and the Commission it may well be that there will be considerable debate about what UK businesses will have to do to satisfy this requirement. Reference to experience in the health and safety context may be valuable in addressing this question.

Environmental effects

BS 7750 requires that procedures be established for examining and assessing the environmental effects of activities, products and services and for compiling a register of those identified as significant. The environmental protection system, which participants in the Eco-audit scheme are obliged to set up, must also include procedures aimed at generating adequate information in order to evaluate the environmental performance of activities. One apparent difference between the two standards, however, is that, whilst BS 7750 makes it clear that the effects arising or likely to arise as a result of past activities must be considered, this does not appear to be the case in relation to the Eco-audit scheme.

This discrepancy may, of course, be of interest to potential participants who have a reluctance to investigate the legacy of past activities, although such reluctance can often ultimately prove to be very costly. In addition, however, outsiders who have an interest in ascertaining the potential environmental liabilities of a business should be aware that any comfort to be drawn from its participation in the Eco-audit scheme should be limited to the potential liabilities which may arise from existing as opposed to previous activities.

The auditing process

The first point to remember is that the nature of BS 7750 and the Eco-audit scheme are different in that BS 7750 is an environmental management system whilst the Eco-audit scheme is an environmental protection system. Consequently, the requirements of the Eco-audit scheme in relation to the conduct of audits are considerably more detailed and wide-ranging than those of BS 7750 which are essentially limited to assessing the effectiveness of the management system adopted. Under the Eco-audit scheme, the objectives of the audit, which must be defined in writing, must include not only assessing the management systems in place, but also "determining compliance with community, national, and local regulations and determining conformity with company policies".

As far as auditors are concerned, BS 7750 requires that auditors have expertise in the relevant disciplines and that they be independent, as far as possible, of the activities being audited. The Eco-audit scheme has a similar requirement in relation to experience in that it requires that auditors have sufficient training and proficiency. It also appears, however, to be an absolute requirement of the Eco-audit scheme that auditors be independent of the activities they audit. It is permissible under both schemes for the auditors to be internal to the participant. Under the Eco-audit scheme, however, the participant has to prepare an environmental statement containing detailed information about the environmental performance of the participating site. This statement must be validated by an external accredited environmental auditor. The accreditation systems, which will be set up by Member States will define the fields in which each accredited expert is authorised to operate. Consequently, where a statement covers complex or wide ranging activities, validation may require the services of a number of auditors having between them the appropriate range of accreditations. So far as BS 7750 is concerned, it is not yet clear how and by whom certification of compliance with the requirements of the scheme will take place. BSI has made it clear, however, that any certification will be compatible with the requirements of the Eco-audit scheme which means, presumably, that it will be undertaken by a third party who has been accredited for this purpose.

The scope of the reviews which follow those audits are similarly different in extent. Under the Eco-audit scheme, management must review regularly the environmental policy, objectives and programmes and, in the light of the latest environmental audit, set new objectives and introduce new measures aimed at improving environmental performance. Under BS 7750, whilst management is required to review the environmental management system at appropriate intervals and take into account the results of audits when conducting the reviews, there does not appear to be any obligation on management to review the environmental policy, objectives or targets. Again, this reflects the fact that the audits conducted under BS 7750 focus primarily on management systems rather than on environmental performance.

Finally, whereas under the Eco-audit scheme, audits have to be conducted at intervals of between one to three years (depending upon whether the participant's management decides that the activities in question have a high, moderate or low environmental impact), BS 7750 appears to give substantial discretion to management as to how often to audit, simply saying that the frequency of audits should be determined "on the basis of the nature and environmental importance of the activity, and the results of previous audits".

Overall, the general impression is that the auditing process under the Eco-audit scheme is likely to be far more time consuming and costly than under BS 7750.

Publicity

Perhaps the most obvious difference between BS 7750 and the Eco-audit scheme – and the difference which may often result in BS 7750 being favoured by businesses – is in relation to the degree of publicity which is required under the two systems.

Whilst BS 7750 requires that the environmental policy be publicly available, it generally leaves it to the organisation concerned to decide upon what other information it will release to the public. For example, although participants must have procedures for responding to queries from relevant interested parties about the environmental effects of their activities, and must have a policy regarding the availability of information, there does not appear to be anything preventing those procedures and policies from being restrictive.

The Eco-audit scheme on the other hand, places great importance on making available information about environmental performance to the public. Indeed, this is one of the stated objectives of the scheme. Participants are specifically required to provide the public with information necessary to understand the potential environmental impact of the participant's activities. They must also prepare an environmental statement for each participating site, following the conduct of an environmental audit at that site. This statement must be written specifically for the public in a concise, non-technical form and must include:

(i) a description of the participant's activities at the site;
(ii) a detailed assessment of all the significant environmental issues of relevance to those activities;
(iii) a summary of the figures on pollutant emissions, waste generation, raw material, energy and water consumption and other significant environmental aspects;
(iv) the participant's environmental policy, programme and specific objectives for the site in question;
(v) an evaluation of the environmental performance of the protection system; and
(vi) a deadline for the submission of the next statement.

This statement must be validated by an accredited environmental auditor who must certify that the statement "gives a true and fair account of the environmental issues of relevance to activities at the site."

The need to produce an environmental statement and, in particular, the need to have it validated by an accredited environmental auditor, is likely to impose significant additional costs on those participating in the Eco-audit scheme as compared to those operating BS 7750. More importantly, however, the need to provide information to the public is likely to add significantly to the external pressures which are brought to bear on an organisation in relation to its environmental performance. It is not difficult to envisage that environmental pressure groups, competitors and others with a vested interest will not hesitate to use any failings or shortcomings identified in an environmental statement to embarrass the organisation concerned. Further, the publication in the environmental statement of the figures for pollutant emissions may well lead to prosecutions by the various environmental enforcement agencies for breaching discharge limits. Not only may the publication of such figures make it easier to identify offenders, but because this information will be available to the general public the pressure on the authorities to take action where it is clear that offences have been committed will be greater than before.

The message for those deciding to participate in the Eco-audit scheme must be to ensure first, that their environmental policy, objectives and targets are specified with great care and second, that the importance of achieving those stated targets is clearly understood by all concerned.

Incentives

It is clear that adopting BS 7750 or participating in the Eco-audit scheme will impose significant additional administrative and financial burdens on business and that those burdens are, for the most part, likely to be greater on those opting for the Eco-audit scheme because of its greater scope. What then are the incentives which will encourage businesses to participate?

As far as BS 7750 is concerned, the standard itself currently provides no specific incentives to encourage participation. Organisations will, therefore, participate because they see sufficient commercial advantages in being able to demonstrate competent management of environmental performance. This may be because they operate in an area in which their environmental performance is an important consideration either for customers or for suppliers or, perhaps, because participation is a requirement of their bankers or insurers. The pilot programme which BSI is currently running is reported to have got off to a very good start with some 300 trade associations and companies involved. The sectors which are reported to have shown the greatest interest so far being chemicals, oil, construction, steel and metal finishing. It would appear, therefore, that there is already a great deal of interest in participating in the scheme without the need for further incentives. BSI is, however, currently considering creating a logo which would be awarded to those who have been certified as operating BS 7750. The question of how and by whom certification will be carried out is still under review. In addition, the Department of Trade and Industry (DTI) is encouraging the participation of small and medium sized enterprises, who may be particularly concerned about whether they have sufficient expertise and resources to participate through its Enterprise Initiative. Through the Initiative, these businesses can, in certain circumstances, recover half of the costs which they incur in engaging a consultant to help in the interpretation of BS 7750, from the DTI.

The Eco-audit scheme offers a specific incentive in the form of a logo showing participation. The logo must be used in conjunction with a statement which identifies which of the company's sites are participating and may not be used on products or packaging or for advertising products. The Eco-audit scheme also allows Member States to establish measures to assist small- and medium-sized enterprises to comply with the requirements of the scheme. Member States will also be allowed to simplify verification and inspection formalities for small- and medium-sized enterprises provided that this is in compliance with European Community law. There has, as yet, been no indication from the Department of the Environment as to whether such assistance will be provided in the United Kingdom but it may well be that similar assistance to that provided under the Enterprise Initiative will be made available to small- and medium-sized participants.

Participation in the Eco-audit scheme should, of course, also carry the same general commercial benefit as registering under BS 7750. Indeed, if anything, these benefits should be greater because of the greater publicity which attaches to the scheme and the greater prestige which is likely to be associated with it. Whereas the scheme is likely to be better known than BS 7750 outside the United Kingdom which may lead to it being preferred by businesses which rely to a significant degree on exports or who seek finance or insurance cover from international institutions, this comment should be qualified having regard to the work toward international standards for management. It will be interesting to see whether the early interest which has been shown in BS 7750 will start to fade once the Eco-audit scheme is up and running for those reasons, or whether the significant additional commitment that will be required to establish a protection rather than a management system will mean that there will be increased interest in BS 7750 as opposed to the Eco-audit scheme.

The EC experience and the context of the Fifth Environmental Action Programme

On 27 March 1992, the European Commission adopted the fifth of its Environmental Action Programmes designed to protect and enhance the quality of the environment in the European Community. This Fifth Programme, (the importance of which is emphasised in Chapter 1) which covers the period 1993 to the year 2000, called "Towards Sustainability" represents a departure from the four previous programmes. For the first time – as a result of the adoption of the Single European Act in 1987 – the Community has a constitutional mandate to take environmental protection measures. Whereas those earlier programmes were very much reactive "end of the pipe" strategies which responded to environmental problems after they had occurred, the Fifth

Environmental Action Programme is a proactive document, based on the thesis of "sustainable development" proposed by the the Brundtland Report. In brief, the thesis put forward is that the causes of environmental degradation should be addressed before the problems become so drastic that they can no longer be ignored. In addition, the Fifth Environmental Action Programme recognises that the full economic benefits of a Single European Market in terms of ongoing growth and efficiency are limited by environmental constraints and that the whole internal market programme could be jeopardised if the natural environment is damaged beyond its limits.

In particular the key strategy behind this programme is based on the fact that environmental damage will never be halted, let alone prevented, if the behavioural patterns of producers and consumers, government and citizens are not altered to take into account the environment much more. Hence, the concept of "shared responsibility" which also underlines this important initiative. The Community Heads of State and Government at the Dublin Summit in June 1990 summarised the thinking behind this Environmental Action Programme. The Communiqué provides the political endorsement to ensure that the words of the programme are implemented in action:

> "We recognise our special responsibility for the environment both to our own citizens and to the wider world. We undertake to intensify our efforts to protect and enhance the natural environment of the Community itself and the world of which it is part. We intend that action by the Community and its Member States will be developed on a co-ordinated basis and on the principles of sustainable development and preventive and precautionary action . . . the objective of such action must be to guarantee the citizens the right to a clean and healthy environment . . . the full achievement of this objective must be a shared responsibility."

The main objective of the Fifth Environmental Action Programme is to raise a public awareness about the problems facing the environment and to change people's attitudes and behaviour so that they become more "environmentally aware". Three principal groups are identified:

(i) Public Authorities (Central and Local Government) have a crucial role to play, not only in enforcing legislation, but also in making planning decisions, and informing and educating the public.
(ii) As huge consumers of materials and producers of waste, public and private enterprise must be both encouraged and obliged to take their responsibilities toward the environment seriously.
(iii) The general public as voter, producer of pollution and consumer of goods and services can exert a great influence on the future quality of the environment.

Among the selected target sectors is industry. In view of the fact that industry is responsible for a major proportion of environmental damage, the new strategy recognises that it is vital to work together with industry to achieve the solution, rather than simply relying on prescriptive measures. The European Community will encourage industry (including small- and medium-sized enterprises) to recognise its responsibility towards the environment by promoting:

- improved resource management;
- consumer confidence in environmental quality and choice; and
- Community standards for production processes and products.

Another intention is that the achievement of state of the art production methods which improve environmental protection should also assist in giving industry within the Community a competitive edge in world markets. It is clear from the earlier discussion that the terms of the Eco-audit Regulation reflect well the idea of "shared responsibility". In its particular reference to the assistance of small- and medium-sized enterprises in this voluntary scheme, it is also clear that encouragement to achieve widespread participation is a priority. The agenda of the Fifth Environmental Action Programme will have increasing relevance for corporate approach to environmental management and will thus have a major influence in the move towards global environmental standards.

The EC experience has shown that the development of regulations to protect the environment is not without problems. One problem that has had to be faced is which out of the two EC principles of "Protection of the Environment", and "Maintenance of the Free Movement of Goods and Services" is to prevail when these two fundamental principles come into conflict.

This question arose in 1989 in the case of Case 302/86 *Re Disposable Beer Cans: EC Commission (United Kingdom intervening)* v *Denmark* [1988] ECR 4607 [1989] 1 CMLR 619. In essence the facts of this case were that it was the practice in Denmark to charge a deposit on the sale of bottles containing beer and soft beverages. The attraction of recovering the deposit was enough to encourage a high percentage of consumers to return the bottles voluntarily and thus the countryside was kept free of empty discarded bottles.

In the mid-1970s, Danish beer manufacturers began to use cans and different shaped bottles. There was thus competition not merely between the drinks themselves but also between the different types of container. Legislation was therefore introduced to ensure that the deposit system continued. Drinks could only be marketed in "returnable" containers (containers for which there was a system of collection and refilling). Containers must also have been approved by the Danish National Agency for the Protection of the Environment. In deciding whether to approve a particular container, the Agency took into account considerations such as whether the system for returning containers was so designed as to ensure that a high proportion of containers were actually returned.

After receiving protests from non-Danish producers of beverages and containers, and from European Associations representing the retail trade, that the containers in which drinks were normally sold could not be used in Denmark, the Commission considered that the Danish provisions were contrary to Article 30 of the Treaty of Rome. The Danish Government amended the relevant law to provide that beverages could be sold in non-approved containers provided that the quantity sold did not exceed 3,000 hectolitres per year per producer or that the beverage was being sold in a container normally used for that product in the country of production in order to test the market in Denmark.

The Commission was not satisfied with this amendment. It considered that the system achieving either re-use or re-cycling was sufficient to achieve the environmental aim and that to limit the volume of the product marketed in non-approved bottles was unjustified. The Commission, therefore, brought proceedings for a declaration that Denmark had failed to fulfil its obligations under Article 30 of the Treaty of Rome. The European Court of Justice found that the conditions limiting the quantity of products which could be marketed by importers to 3,000 hectolitres per producer per annum did constitute a breach of Article 30.

However, the importance of the case lies in the re-statement by the Court that:

> "The protection of the environment has already been considered by the Court in Case 240/83 *Association de Defense des Bruleurs d'huiles Usagées* [1985] ECR 531 as 'one of the essential objectives of the Community' which may, as such, justify certain restrictions on the principle of the free movement of goods. Furthermore, this assessment is confirmed by the Single European Act."

The Court went on to say that:

> "in view of what has been said it must be concluded that the protection of the environment is a mandatory requirement which may limit the application of Article 30 of the Treaty."

It is thus clear that in appropriate circumstances where there is a conflict within the European Community between the principle of protection of the environment and the principle of free movement of goods, the principle of protection of the environment will prevail.

A further facet of the potential tension between the principle of protection of the environment and the principle of free movement of goods lies in the question of whether standards that are introduced in the European Community are absolute standards which are not to be exceeded or whether they are minimum standards. This issue has been brought into sharp focus in the context of packaging materials in Germany where the German authorities have taken the standards imposed by EC Directives on re-cyclable packaging as minimum standards and have significantly exceeded them, thereby creating conditions of virtual impossibility of access into the German market for non-domestic producers of packaging. Concomitant with such virtual impossibility of access runs the argument of impediment to free movement of goods. It is likely that the 1990s will see significant litigation in the European Court of Justice on issues of this nature.

Freedom of access to environmental information

Bearing in mind the thrust of the Fifth Environmental Action Programme and its furtherance of the "polluter pays" principle, as well as the development of the citizens' rights approach, Canal Directive 90/313/EEC on Freedom of

Access to Environmental Information, which was published on 7 June 1990, is a clear demonstration of the trends in European environmental protection. Coupled with the decision on 7 May 1990 to set up a European Environment Agency (EEA), the commitment to involve the public in environmental issues is extremely evident. It is clear that the Commission regards the increasing availability to the public of environmental information as an integral part of its environmental protection strategy. The fact that this Directive is referred to in the Preamble to the Eco-audit Regulation provides further evidence of its commitment.

The Directive requires Member States to ensure that information relating to the environment held by public authorities is made available to any person (natural or legal) at their request. The effect of this Directive was enhanced by legislation establishing the EEA (whose location is still be to agreed) and the European Environment Information and Observation Network through which information relating to the environment can be collected and disseminated to the public. The joint effects will be an improved EC environmental monitoring system, with an increase in the quantity and quality of environmental information. In turn, these effects will increase public participation in the procedures to control pollution, including the enforcement of environmental laws. The Directive is also intended to ensure the removal of disparities in Member State laws which cause unequal conditions of competition.

For the purposes of the Directive the "environmental information" is:

> "any available information in written, visual, aural or database form held by public authorities on the state of water, air, soil, fauna, flora, land and natural sites, and on activities (including those which give rise to nuisances such as noise) or measures adversely affecting, or likely so to affect these, and on activities or measures designed to protect these, including administrative measures and environmental management programmes."

Anyone in the European Community would be entitled to information, without the need to show an interest, at a reasonable cost, within two months of the request for information. While there are exceptions to the various requirements for information to be made available, for example, where publication of the information would affect national security or commercial confidentiality, industry's environmental performance is undoubtedly going to be open to considerably greater public scrutiny than before. The due date for compliance by Member States is 31 December 1992. The United Kingdom has done better than some of its EC partners in meeting this deadline by implementing the Directive by means of the Environmental Information Regulations 1992 (SI 1992/3240). However, the list of relevant persons (although not entirely clear, this would appear to mean public bodies which have environmental responsibilities and bodies controlled by them) to which the Regulation applies is not provided; as such some uncertainty exists, by way of example, as to whether privatised utilities will need to comply. British Gas has said that it is a relevant person and Friends of the Earth is planning to test the Regulation soon. The trend is clear and it is recommended that industry "gears up" accordingly.

Strict civil liability for waste

Over two years ago the Commission published a proposal for a Directive which would make a producer of waste liable under civil law for damage to person or property and for injury to the environment caused by the waste, irrespective of any fault on the part of the waste producer. Discussions on the proposals contained in the Directive have included suggestions that it should contain a financial ceiling on liability and a "state of the art" defence. The proposed Directive has proved both political and controversial, consideration being given to the massive ramifications for business of the US "Superfund" strict liability "deepest pocket" regime, as well as the increased application of the concept of harm or impairment to the environment.

While it is experiencing a lengthy legislative passage, should this Directive be adopted, it would almost certainly have consequences for the proposed Eco-audit scheme. In particular, because the producer of waste is made civilly liable irrespective of fault; and environmental pressure groups are given standing to sue for injury caused to the environment by waste irrespective of whether they themselves have been injured.

The remedies available include the right to obtain the "reinstatement of the environment" and the reimbursement of any costs the organisation has legally incurred, presumably including reimbursement of legal fees. If one then considers the Eco-audit scheme's emphasis upon greater public access to information, it poses questions as to how far companies will wish to participate in such a scheme – whether or not there is a limit on liability to incidents occurring after the Directive's effective date. For instance, in practice, a company will have to assume – and prepare for the costs of – public disclosure of any deficiencies revealed by the Eco-audit because, if left uncorrected, this could lead to action by either an enforcement agency or environmental organisation. It is clear that progress in the Civil Liability Directive could have major ramifications for the Eco-audit scheme.

The eco-labelling scheme

In February 1991, the European Commission submitted a proposal for the establishment of an eco-labelling scheme whereby eco-labels could be awarded to products that are judged, by reference to specified criteria, to have a less significant impact on the environment than other similar products. This initiative is referred to in Chapter 1. The idea is that the criteria will cover all significant environmental effects of products throughout their life cycle, from manufacturing to disposal: the so-called "cradle to grave" approach. They might, for instance, cover how much energy is involved in making and using the product and how easy it is to re-cycle or re-use. (The recent Energy Label initiative should also be considered by businesses in their approach to environmental manufacture.) Such criteria will be valid in all Member States: the scheme therefore provides for an official and independent seal of approval that is European-wide. Other key features of the scheme are that it will be voluntary and self-financing, and that it will not cover food, drink and pharmaceuticals.

In 1991 the Community set up *ad hoc* working groups to undertake pilot studies to test "cradle to grave" assessment for the scheme. Certain test products have been under scrutiny in Member States as part of this pilot. The first product categories have been announced and the determining criteria have been made the responsibility of Member States as follows:

Packaging	Italy
Kitchen towels	Denmark
Toilet tissue	Denmark
Photocopy paper	Denmark
Writing paper	Denmark
Household cleaning products	Germany
Laundry detergents	Germany
Dishwasher detergents	Germany
Paints and varnishes	France
Insulating materials	Denmark
Washing machines and dishwashers	United Kingdom
Soil improvers	United Kingdom
Hairsprays	United Kingdom

Additional product groups are batteries, domestic heating systems, refrigerators, building materials and textiles.

The official text of the Regulation was finally published in April this year (OJ 1992 L99/1) following its agreement by the Environment Ministers in December 1991. The Regulation was adopted on 23 March 1992, which also appears to be the date of its entry into force. Member States have six months after this date to nominate the "competent bodies" which will carry out the tasks devolved to the national level under the scheme. Such "competent bodies" will: propose product groups for inclusion in the scheme; assist in the development of criteria for products; assess individual applications for a label; conclude contracts with successful applicants. Legislation establishing the scheme had to be in place to enable the scheme to become operational as of Autumn 1992 to allow Member States sufficient time to make the necessary preparations for its implementation.

The "competent bodies" will have an important role in respect of the development of the determining criteria: they will work up the proposals for product group criteria in consultation with industry, commerce, consumer and environmental organisations and other interested parties. As the criteria will be changed to keep abreast of new technologies and processes, it is to be anticipated that the criteria will be forever tightening. The judging criteria will be amended to achieve the highest possible environmental performance. Probably the normal period of validity for criteria will be three years: this will vary, however, depending upon the rate of technological development in the various product categories.

The flow diagram below indicates the procedure for establishing product groups and criteria:

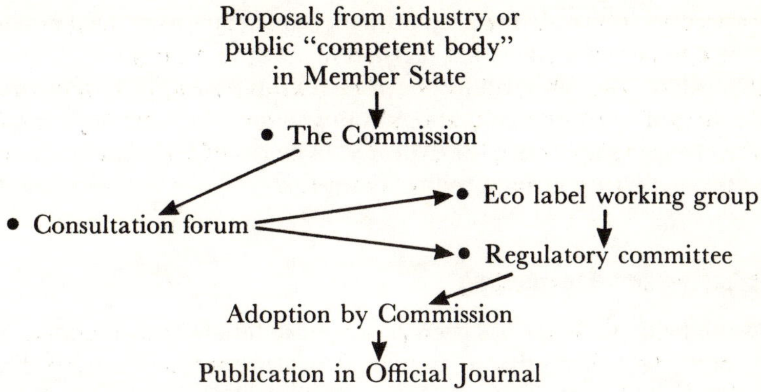

The Regulatory Committee consists of the Member States and the Commission.

The Regulation envisages that the "competent body" of each Member State will act as a first screening body of application for eco-labels. Together with their application, applicants will pay a non-returnable fee for the initial assessment and a yearly licence fee thereafter. The fees are still to be determined: however, they are likely to be modest since the organisers want to ensure that this cost does not mean a price barrier for small businesses. If it is satisfied that a certain product satisfies the eco-criteria applicable to it, the "competent body" will inform the Commission that it proposes to award a label, and the Commission, in turn, will inform the "competent bodies" of other Member States. If neither the Commission nor any Member State objects within 30 days, the label is awarded. If there is an objection – which must be reasoned – a further 15 days are allowed to resolve the matter. Unresolved objections are submitted to the Regulatory Committee which makes a final decision.

The flow diagram below illustrates the procedure for awarding the eco-label to individual products:

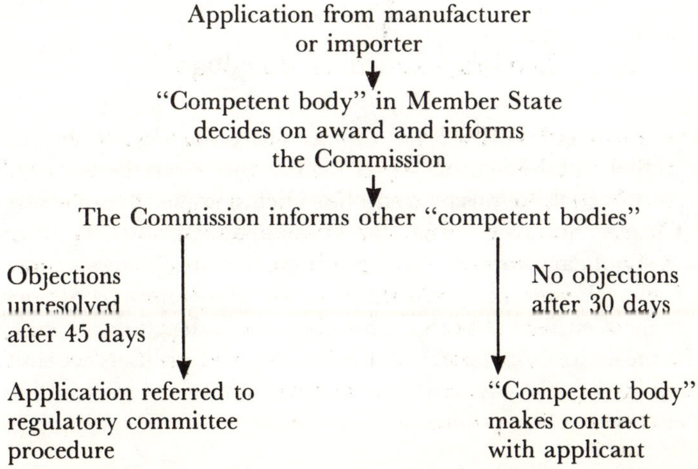

The procedure is intended to keep bureaucratic delays to a minimum: the aim is that the award for a label for a product in an agreed category will take three months or less. The label will be issued on a simple pass/fail basis without any graduations of environmental acceptability, such as a "star" system. Labels, and the licence to use them, are expected to be issued for a maximum of three years, depending upon the product category.

Eco-labelling and Eco-auditing

The eco-labelling scheme has been perceived to be a matter of priority in view of the misleading claims that abound in the green consumer market. The logo, in the form of a flower, is intended to distinguish the truly more environmentally benign products from the rest. In addition to providing authorative guidance to consumers, the scheme aims to encourage production of, and trade in, the eco-labelled products. Like the Eco-audit scheme, this scheme seeks to provide an incentive to participants to improve their environmental performance. Both schemes have similar purposes and have a similar vehicle – a logo with official EC status – to inform the public about a company's environmental soundness.

Eco-label award scheme logo

The eco-label legislation will not impose any obligation on any person to apply for a label and the fact that a product does not merit the award of a label will not result in prohibitions or restrictions being imposed on the sale of that product. Clearly, however, businesses whose products do carry an eco-label will have a significant commercial advantage over their competitors.

Since both schemes are voluntary a company can choose which to participate in, if either (or both). This choice will depend on a cost/benefit analysis for the company; of particular relevance are the differences in the level of environmental protection required and the amount of information which must be disclosed to the public:

- Unlike the Eco-audit scheme, the eco-label scheme does not require an extensive environmental protection system: as it is based on product criteria there is no need for the applicant to put in place an environmental protection system of the comprehensive nature envisaged by the Eco-auditing scheme.
- As regards public disclosure, whereas authorisation to use an Eco-audit logo is based upon a "fair and accurate" public accounting of a company's environmental performance, the eco-label involves much less public disclosure: all information regarding a product is treated as confidential until a decision is made to award the label (there is no comparable provision in the Eco-audit proposal). Even after the eco-label is awarded the disclosure is not extensive, for example, little or no information need be produced regarding the process by which an approved product is manufactured.
- In contrast to the Eco-audit scheme where public disclosure is focused on performance, *i.e.* process, public disclosure is focused entirely upon the environmental soundness of the product in the eco-label scheme.

Given the fewer demands on a company, the less onerous public disclosure and the benefit of an eco-label, it may be that this may be considered more valuable, especially as the eco-label can be used in conjunction with the advertisement of a specific product. We have yet to see how far the use of the eco-label will in fact affect the implementation of and popularity of the Eco-audit.

Towards a world standard

Assuming that BS 7750 (like its sister, BS 5750) is adopted as a European Standard, the next step will be to work towards the adoption of a world standard on environmental management systems. The creation of a world standard will be moving into hitherto untrodden territory. The difficulties in securing adoption of such a world standard, arising from social, cultural and economic differences are likely to be considerable. However, these will, in all probability, be regarded as being outweighed by the trans-boundary effects of the environmental problems that face the world and the importance of resolving those problems.

BS 7750 is already having a significant impact on the international stage. The United Kingdom is the convenor of an International Standards Organisation (ISO)/International Electro-technical Commission (IEC) working group on environmental management that is now preparing a document that could become a draft International Standard on the subject. The working group is set up under the aegis of the Strategic Action Group on the Environment (SAGE) which was established jointly by the ISO and the IEC last year. The group started the process of designing a brief for an

International Standard from scratch at a meeting in May 1991. This process enabled all the representatives to cite specific characteristics that the document should have, as well as to compile a small list of selected documents to be used as source material. BS 7750 has been included on that list by the working group, as it is the world's only standard on the subject of environmental management systems. A draft document was submitted to the representatives of the working group in June 1992. The new technical environmental committee (which matched that of BSI) was based upon an official proposal to produce international standards on:

- environmental management systems;
- environmental auditing;
- environmental labelling;
- environmental performance evaluation;
- an industrial mobility plan for developing industrial standards; and
- life cycle analysis.

The intention is to produce an ISO Standard some one or two years after the revision of the BS 7750 Standard (which is likely at the end of 1993) and the establishment of certification bodies in 1994, provided that there is agreement on the production of an International Standard. It is probable that this will involve a systems approach which embraces an environmental management system, a health and safety management system and a quality management system.

Following the qualified success of the United Nations Conference on the Environment and Development (UNCED) 1992 Conference in Rio de Janeiro, the agency that is likely to be best placed to promote the development and implementation of world-wide environmental standards (including world standards on environmental management systems) may be the United Nations. If the developing momentum to establish an International Environmental Agency as a permanent organ of the United Nations achieves its goal, it may be that such an International Environmental Agency would be the most appropriate organ of the United Nations to develop and establish world-wide environmental management systems standards.

It is appropriate, therefore, briefly to trace the developing movement in favour of the establishment of an International Environmental Agency and of an International Court for the Environment. In May 1992 the International Court of the Environment Foundation (ICEF) was constituted in Rome with scientific backing from a number of leading Italian organisations and enterprises. The Foundation is a non-profit organisation set up for the purposes of promoting continuous action regarding the International Court of the Environment and public awareness about it: see, on the subject, Amedeo Postiglione *The Global Village Without Regulations: Ethical, Economical, Social and Legal Motivations for an International Court of the Environment* (1992) 75.

The constitution of the Foundation is the logical outcome of the work done by an original organising committee which was founded in 1988 with the specific aim of ensuring that the problem of the institution of an International

Court of the Environment be discussed at the UNCED 1992 Conference. The Foundation advocates the establishment of an International Court of the Environment along lines set out in a draft statute for such a Court which it has produced. This draft statute is divided into three parts. The first summarises various principles. The second establishes the International Environmental Agency as a new permanent organ within the United Nations which will be responsible for environmental control and monitoring, promoting research, planning initiatives related to prevention and environmental restoration, management of the world environmental fund, the setting of standards, and the publication of a triennial report on the world ecological revolution. The third part of the draft statue deals with the establishment of the International Court of the Environment as a new permanent organ of the United Nations (with the task of resolving disputes about the environment which, due to their nature, touch vital interests within the international community and with powers for taking urgent measures in the event of environmental disasters or of situations that seriously endanger the environment). It is envisaged that this body would be accessible not only to the various nation states of the world but also to individuals and to non-governmental organisations (*Ibid* 49).

It is clear that the development of such an International Court of the Environment is predicated upon the development of further aspects of international law and acceptance by nation states of derogation from state sovereignty. As a concept of international law, sovereignty is divided into three major aspects: external, internal and territorial sovereignty. All these aspects are fundamental to state sovereignty and sovereignty in turn has probably been considered up to the present time the most fundamental principle of international law, as it forms the basis of international relations and is the concept from which many other principles of international law stem. However, neither the United Nations Charter nor international customary law recognise absolute sovereignty as an inviolable tenet of international law. Further, the trans-boundary inter-related nature of environmental problems no longer admits of a "traditional" approach to international law based upon the sovereignty of nations. The exigencies of the environmental crisis facing the world are such that Nation States will have to derogate from their sovereignty in order that effective means of dealing with environmental crises can be established for the benefit of all.

One of the most important and authoritative declarations regarding the need for an International Court of the Environment has come from the European Parliament in Resolution A3 – OJ 1992 C67/142 on EC participation in the UNCED. This clearly states that the European Parliament:

> "14 Considers the existing international law too weak in terms of both monitoring and penalties and the enforcement of penalties: therefore advocates the tightening up of international law world-wide and calls for the setting up of an International Environmental Court with world-wide jurisdiction, either at the International Court of Justice at the Hague or at the United Nations Offices in New York."

In recent years, industry's relationship to the environment has become a primary concern to society in general. In response to this concern, considerable amounts of UK and EC legislation have been enacted and it is likely that there will be significantly more to come in the near future. If industry is to achieve compliance with this legislation and in particular the UK Environmental Protection Act 1990, it will have to operate within the constraints of increased financial costs (and benefits!) and environmental liabilities.

Environmental protection has become a major issue for society and is attracting increased interest from the public, politicians and industrialists. There is little doubt that environmental protection will be on the agenda of companies, trades unions, banks and insurance firms for many years to come. The level of environmental protection awareness is increasing and most industrialists recognise this. As a former chairman of British Coal clearly stated:

> "Management ignores environmental issues at its own, literal peril. A knowledge of the issues involved is a vital part of the armoury of any manager attempting to carry out his or her day-to-day functions in the last quarter of the twentieth century."

Indeed, there can be no doubt that if environmental protection issues are ignored then many businesses, industries and possibly some economies will flounder upon the "polluted rocky shores" of ecological disaster.

The 1980s saw a growing concern throughout Scandinavia, North America and Europe with respect to environmental issues. Increased awareness of environmental problems has led to the introduction of stringent new UK and EC legislation. These Regulations and Directives require rigorous assessment, reporting, implementation and monitoring to ensure compliance.

In the short-term the pursuit of environmental excellence, by companies and/or countries is going to "cost". To achieve this goal many techniques and methods will have to be employed. However, to ensure that all parties concerned are able to compete in this area, it is necessary to have some guidelines and/or rules, which are based upon clearly defined objectives and are not punitive to any of the participating parties.

By using proven and validated environmental standards a consistent and reproducible approach to the identification of environmental concerns and potential risks can be developed. Also, by using such standards, the data derived can be subject to rigorous and stringent evaluation. Thus, greater confidence can be placed in the findings than can be placed upon data derived from studies where standards are not employed.

However, one of the most important features of the application of environmental standards is that of enabling companies and/or countries to compete in the world markets on the basis of a "level environmental field". This is particularly relevant at this point in time due to the appearance of the European Single Market. This should, in theory, allow all Member States of the European Community to compete on equal basis with each other. Thus, harmonisation of trade etc should not be inhibited by environmental strictures. In practice, however, such a situation is unlikely to be achieved in the near future. Additionally, the cost to achieve this goal will be very high, and companies within the European Community (and those that wish to trade with EC countries) will find that the "level playing field" is perhaps not as level as first thought.

A result of this inequality may be the creation of situations where the ability of companies or countries to perform in the market place is impaired or, in extreme cases, non-existent. Thus, in the case of developing countries that have to compete against developed ones, compliance with environmental standards may have severe consequences for their economic, social and political growth.

CONTENTS OF CHAPTER 5

Total Quality Management and the Environment

Chapter 5
Total Quality Management and the Environment

Total Quality Management or TQM is one of the "buzz" words of the 1990s. Cynics have criticised the acronym as a passing fad but, while the terminology may have changed by tomorrow, it is becoming very clear that the philosophy it represents is here to stay. This chapter will take a look at the philosophy of quality management which lies behind the TQM jargon, some of those who have been pioneers in the bid for total quality, and what the environmental implications of TQM might be, given the growing pressure on businesses to improve their environmental performance.

The philosophy of TQM

A competitive edge

In the world of business today it has become impossible to ignore the overwhelming drive towards greater efficiency, reduced costs and increased productivity. Quality and "value for money" are the two prime movers for the potential customer deciding where to place his business, and consequently companies are beginning to recognise quality as a major strategic issue. This change in philosophy is having an impact on every business and all those who depend on them. It is becoming essential to have the ability to compete in a market which is increasingly *global*, particularly as we move towards a more unified Europe, and advances in technology and communications make the world market more easily accessible.

In general terms, all companies compete on three principal factors: quality, delivery and price. Although much of the Western world is currently in a state of economic recession, there are few who would deny that quality is the most important of the three. The present commercial environment is such that companies must strive for competitive advantage to hold on to market share, let alone increase it. Consumers now place a higher value on quality than on loyalty, and price is no longer the major determining factor in consumer choice.

The quality business

The concept of quality in business is not really a new one; emerging with the era of mass production, its development can be traced through the periods of mass inspection, quality control, and quality assurance through to quality management. From the mid-1980s, the concept to grapple with has been Total Quality Management. TQM, used properly within an organisation, has been described as a circle embracing the whole organisation and symbolising continuous feedback and improvement.

TQM is a method of business management designed to improve the effectiveness, flexibility and competitiveness of any given organisation, from the top to the bottom. It must embrace the *whole* organisation: it involves each department, each activity, and each person, at every level. It extends far beyond the production process or customer service, because in order that a company may be truly effective, all its constituent parts must work in harmony, since each person and each activity affects and is affected by others. TQM is a complete and continuous process, hence the image of a circle of constant feedback and improvement, not the straight line of conventional business process analysis. It is a method of management which applies equally to both the service industries and to the manufacturing and processing sector.

The objective of TQM is to satisfy customer requirements as efficiently and profitably as possible. The basic philosophy has been described as "doing the right things right, first time, every time", (Munro-Faure "Implementing Total Quality Management"). TQM has become the term to be applied to any form of customer-focused corporate culture which concentrates on continual improvement, but there are certain fundamental ingredients which are necessary for its successful application. TQM begins by identifying all a company's customers, in the broadest sense of the word, one of whom should be the environment. Consideration for the environment can be built into a total quality programme in exactly the same way as consideration for the people who pay the bills. Identifying an organisation's customers, the requirements of those customers, and how those requirements will be met, form the core of the total quality approach. To succeed in this approach a company will need in place certain key elements: a good quality management system, statistical process control (SPC), and teamwork – all of which will be discussed in more detail in the following sections.

Implementing TQM

Quality chains

Perhaps one of the most widely-accepted definitions of quality is "meeting the customer's requirements", but the pioneers of TQM have suggested that it goes beyond merely meeting requirements; the customer should be *delighted* by the quality of service the organisation is providing. All businesses involve a series or chain of processes to produce the product, or to provide the service the customer requires. These chains stretch beyond the organisation itself, so they not only incorporate internal departments and offices, but also chains of external suppliers, whether they be banks, retail outfits, manufacturing concerns, hospitals or education establishments.

Quality must be achieved and maintained throughout the internal and external supply chains. Yet the quality chain can be broken at any point, by one individual or a single piece of equipment failing to meet the requirements of the immediate customer. This might be something as low-key as a photocopier failing to produce legible copies, so that internal record-keeping is impaired, or it might be a secretary who types an invoice incorrectly and thus causes potential damage to the organisation's relationship with a customer. It is, therefore, important for all employees within an organisation to understand who are their immediate "customers" and "suppliers", in order to appreciate their role in the quality chain. Involving everybody in the quality process is essential for the successful implementation of TQM.

Management systems are the means by which customer/supplier quality chains are forged, strengthened and maintained. They are essential to establish the desired objectives of the quality strategy, to make quality part of every activity within the business are the responsibility of every employee. Quality systems have to be reviewed regularly against corporate objectives and the quality strategy. However, it is important to understand that management systems are only a facilitator in the process of achieving TQM; companies who have a management system in place will not automatically qualify for TQM status.

Commitment

There are a number of other conditions which must be satisfied before the TQM implementation process can begin. It is fundamental that the management is totally committed to the quality revolution and that it agrees and understands what has brought about the need to initiate a TQM programme. There must be clear leadership from the top to show the workforce that TQM is more than a management whim, as it will not be possible to implement a quality programme without their support. It must be decided who will have responsibility for managing day-to-day activities, and who will provide direction and influence commitment. These are key decisions because they will have a significant bearing on the initial reaction and subsequent involvement of senior management and the workforce.

Targets and benchmarks

Management needs to be clear about what it is trying to achieve and it should establish well-defined targets. It is necessary to design a quality programme which is tailored to suit the culture, personality and customers of the company, based on a sound understanding of where that company intends to be in the future; not just in two years' time, but in five or ten years' time. The programme should address issues of quality improvement in relation to the processes involved with running the business, such as internal relationships between departments, individuals and technology, as well as external relationships between the company and its customers and suppliers. In other words, the quality programme needs to run in harness with the company's long-term business plan.

The quality process must be consistent at all levels, and the emphasis should be on continuous improvement, with a scheme to measure quality performance against targeted criteria designed to indicate progress and maturity. Performance measurement is extremely important in order to facilitate improvements, as it enables a company to focus its efforts and utilise positive and negative feedback about the changes being introduced. Consequently, auditing is a fundamental aspect of the quality management process.

Chapter 1 provides an overview of the concept of "Eco-auditing" and takes a look at the issues which need to be addressed when trying to define an "audit". Auditing for the purpose of performance measurement clearly involves a broader brief than that for an Eco-audit, but the underlying principles are not dissimilar. Chapter 2 concludes that there are certain ground rules which characterise an Eco-audit: it must be systematic, both in terms of methodology and scope; it should not simply be a question of confirming compliance with existing legislation, but should look at the broader issues such as site health and safety and employment policy. These guide-lines are clearly relevant in the context of performance measurement.

Customer care

It is very important that while a company is addressing the implementation of its quality programme, it does not lose sight of the ultimate motivation for the process – the customer. Is there a culture of listening to customers and actively considering their complaints and suggestions? TQM means that a company's entire business operation must be geared towards customer satisfaction, or even customer delight, both within and beyond the company, because that in turn leads to maximum commercial success, which, for most of us, is what it is all about. It is in this context that service level agreements are gaining increasing currency and importance.

The cost of quality

Consumers have ever-increasing expectations, while companies do not have limitless funds to invest in improving quality. While nobody would deny that implementing a quality management system of any type does involve a certain degree of investment, it is very important to view that investment as a means to an end; the long-term aim being increased profits and expanding market share.

Increased profits

As the quality of an operation improves, through the implementation of TQM, significant amounts of money can actually be saved through the minimisation of failure and detection costs. If instances of defective quality are reduced then so too are the "behind-the-scenes" operations which are dedicated to dealing with problems, errors and wastage. Costs are then saved all the way down the line as delivery performance benefits from increased output and higher productivity.

There should also be considerably less waste produced when a total quality system is in operation at all levels, with every member of an organisation understanding their part of the quality process, and the importance of their role in achieving a successful end-product. Reduced wastage and minimisation of failure will mean, in addition, increased energy efficiency and therefore greater cost savings. What is more, performance failures, such as poor quality and non-achievement, have a habit of multiplying so that there is a knock-on effect felt throughout the production chain. Most people in business spend a considerable amount of their time trying to counteract performance failures: looking for things, checking things, chasing, rectifying problems, redoing tasks and apologising to customers.

It has been estimated that about one-third of all effort expended in the course of business in Britain is wasted due to the snowball effect of performance failures. If it were possible to represent that statistic in financial terms, the cost to British industry, and the overall effect on the economy, would be staggering. Very few organisations actually have a reliable way of measuring the cost of internal failures in terms of lost profit and competitiveness, yet clearly the potential for saving money and gaining market share in this area is considerable.

Expanding market share

In addition to boosting profits by minimising performance failure and wastage, TQM should give a company a competitive edge in its market place. If implementation of the TQM process is successful and customers are genuinely delighted, then a company will have enhanced its reputation. Good public relations should place a company in a position of strength from which to take on its competitors. If a company is consistent in its aims to get each job done right first time, it can reduce its energies in fields such as complaints administration, product returns, product recalls and product replacement, and concentrate on administration of its quality assurance system, training of employees and product development.

However, TQM is not about cost-cutting nor is it a device simply for boosting productivity. While a successful programme will doubtless result in benefits of this kind, they must not be the motivation for a quality system. If a company focuses on short-term financial benefits, rather than on long-term achievement of a total quality operation whose ultimate aim is satisfaction, or delight, of its customers, then it will never achieve its maximum business potential.

The pioneers of TQM

The gurus

The quality movement can probably be said to have been started in the early 1930s by WA Shewart, who, in 1931, provided a definition of the concept of Total Quality Control. He considered how quality control might be applied to a process and measured. He started from the premise that variability in mass production is inevitable, given the differences between the abilities of employees and performance of machinery and parts. However, he argued that use of statistical and probability techniques would allow variability to be better anticipated, and consequently monitored and controlled. He is recognised as the originator of SPC techniques, which still play an important role today for companies seeking to expand their market share, having embraced the quality ethos. (See further Dr Mohammed Zairi *Total Quality Management for Engineers*.)

Quality as a business concept has since undergone various evolutionary changes. Well-known names associated with its development include W Edward Deming, an American scientist whose interest in the application of statistical techniques resulted in an invitation by the Japanese in 1950 to give a series of seminars on the use of charts and statistical techniques for quality control. He also encouraged the Japanese to go beyond the utilisation of statistics and aim for continuous quality improvement using modern consumer research, customer surveys, and other techniques, to keep abreast of market changes. Deming believes strongly that a quality revolution should be led by the management, and that it is the management's responsibility to create a positive working environment, so that workers enjoy the work they perform; working with a sense of purpose which enables them to develop pride in what they do. (*Ibid.*)

Another pioneer in the bid for Total Quality Management has been Joseph M Juran. Born in Romania, but brought up in America, Juran is viewed by many as being the co-founder, together with Deming, of the Quality movement. In 1951 he drew attention to the relevance of quality in relation to reducing costs and raising standards in his Quality Control Handbook. He too was invited to speak in Japan at the same time as Deming, on the subject of planning, organising and managing quality programmes. (*Ibid.*)

The Japanese example

In terms of pioneering TQM, the Japanese are recognised as leading the field. While in the West it is only in the past few years that managers have begun to recognise the relevance of quality as a major strategic issue, the Japanese quality revolution began in the mid 1940s. This time was a crucial period for the Japanese who were concerned to start afresh and rebuild Japan after the Second World War. Japanese products were not renowned for their reliability and quality, indeed they were often noted as being cheap and shoddy. This period also coincided with the early teachings of Deming and Juran who, together with the Japanese Government, played an important part in persuading Japanese companies that concentrating on quality would provide their best chance of competing in international markets. It was this successful shift in manufacturing strategy which lead the Japanese to embrace the philosophy of TQM.

In the past 20 years, the Japanese have moved into a position of international strength, out-gunning many sectors of Western industry, notably the automotive and electronics industries. Their success lies in the fact that Japanese companies now have a proven track record for quality and reliability at a competitive price. Through the reputation which they have established, the Japanese have effectively raised customer expectations to the extent that, in many cases, competitors have simply been squeezed out of the market. Customers now *expect* quality and reliability at a price they can afford, because the Japanese have demonstrated that such standards are consistently feasible. It is not just the manufacturing sectors which have been dominated by the Japanese; in the last decade they have concentrated on quality in the service sectors, and already have a dominant position in financial services, owning seven of the world's ten largest commercial banks.

How the West was won

Meanwhile, in the West, companies are finally waking up to the fact that quality is the key to success. As the economy weakens and squeezes profits, companies are finding that providing better products or services is the surest path to success. "The real weak ones are going to disappear in this economic downturn", says Don A Rice, an American distribution expert who has noticed a recent upturn in bookings for his $1,995 Quality Improvement seminars at Texas A & M University, (*Business Week*, Issue 3195 (McGraw-Hill, Inc)). It seems that when markets are vulnerable, the only salvation lies in doing things better. One US company to have proved the point is Wallace, a $79 million company who distribute pipes, valves and fittings for the oil and chemical industries. Despite the reduced Gulf Coast market in the past few years, Wallace are reported to have boosted sales by more than 69% since 1987, increased their operating profits by more than sevenfold, and raised their market share from 10.4% to 18%, (*ibid*).

In 1990 Wallace was the first wholesale company to win the prestigious Malcolm Baldridge National Quality Award for small businesses, which is the American answer to the Japanese Deming Prize. Curt W Reimann, the Director of the Baldridge Award Programme, says that Wallace "have a passion for the Total Quality Management System" and goes on to suggest that the key to their success has been to encourage worker motivation through training. Since 1987 Wallace has spent $2 million on developing an operation within which its 28 employees work as a team. By creating an atmosphere where everyone's opinion counts, and the need for a top quality performance from everyone is understood to be paramount, quality improvement is dramatic. In 1987, only three out of four deliveries arrived on time; by 1990 Wallace was hitting 92% of target deliveries, and was aiming for 98% within the next year, (*ibid*).

Commercial pressure

Quality improvement is not always the result of self-motivation within a company; buyers are pushing wholesalers to be more quality-conscious. In an economic climate which is forcing everyone to try that much harder to do business, major customers are putting pressure on their suppliers for quality improvement by saying, in effect, "if you want to supply my company, you have to have a quality management programme in place." In this way, the drive for quality is forced down the supply chain, having an impact upon increasing numbers of companies.

TQM – a success story?

Four decades after the drugs company, Tanabe Seiyaku, won Japan's first Deming prize, many Western companies are still looking to Japan in an attempt to develop an understanding of the TQM philosophy, and to try to fathom whether it *is* the key to Japan's phenomenal success in international commerce. Some Western companies, such as America's Motorola, no longer need to be convinced. Motorola claims that TQM techniques have cut somewhere in the region of $700 million from the company's manufacturing costs over the past five years. Yet there is growing evidence that not all converts to the quality movement have so much confidence in their policies.

Disappointment

Arthur D Little, consultants who regularly advise clients on implementing quality programmes, earlier this year carried out a survey of 500 American manufacturing and service companies. Results of the survey showed that only a third of those questioned felt that their quality programme was having a *real* impact on their competitiveness. A similar study, by consultants AT Kearney, carried out recently on 100 British companies, showed that only a fifth of those questioned believed that their investment in quality programmes had yielded "tangible results". It would appear that of those Western companies who have had a quality programme in operation for more than two years, two-thirds have allowed their commitment to wane as hoped-for results did not become reality (*The Economist Newspaper* Vol 323, Issue 7755.)

Somewhere in the region of 75% of British and American companies claim to have some form of quality programme in place. If the statistics are to be believed – cynics might say that many of those companies will not have quality programmes which can compare with those operated by Japanese companies – why are Western companies finding quality such a disappointment ?

Where the West is going wrong

One of the reasons stems from the relative inexperience of Western companies. Even those companies considered to be among the most advanced in terms of TQM, have had barely 10 years' experience. This cannot compare with Japanese companies such as Nippondenso, Honda and Nissan who were among the first pioneers of TQM, inspired by Deming and Juran in the 1950s.

Inexperience is not the only stumbling block for businesses in the West; many companies are failing to channel their energies effectively, unable to grasp that the focus of their efforts should be the customer. There is a tendency to concentrate on processes rather than the end result – the theory being that getting the manufacturing or service process right leads to a quality product. However, more often than not, the customer gets overlooked and then the whole object of quality improvement is lost. The customer must be recognised as being at the hub of the quality programme wheel.

Systems which concentrate primarily on the quality of processes are not, in reality, TQM systems, but a form of quality assurance. BS 5750, the British Standards Institution's (BSI) quality management system is an example of a management system which aims to achieve quality assurance, or "total quality control". While it has some of the makings of TQM in that it is concerned with continuous quality improvement, it is less proactive than TQM. Total quality *control* is based upon a system of checking and maintaining product quality, rather than being an integral part of a firm's culture, relying upon chains of quality which run unbroken through the firm and extend beyond.

Many Western companies have fallen into the trap of implementing superficial quality programmes which concentrate primarily on reaping the public relations and marketing benefits of TQM. Companies need to understand that it will take more than a fresh marketing initiative to convince customers that corporate thinking has undergone a revolutionary change. Quality must be in the eye of the customer; simply dipping a toe in the water will not convince – total immersion is required.

Similarly, some companies imagine that TQM involves investing heavily in a Quality Department, whose employees will be responsible for implementation of a quality programme. Quality must be in a company's blood; it needs to be regarded as "how to do business". Just as a company does not have a Department to deal with "doing business", it does not necessarily require a Quality Department.

Perhaps the most fundamental error made by Western companies is their impatience for a "quick fix"; TQM is about long-term success. While TQM *can* result in some fairly swift cost-savings initially, the long-term effects are certain to represent the greatest rewards. People take time to evaluate and respond to changes, and if their own values are challenged by management controls they may take a long time to adapt. Any significant shift in an organisation's culture, predetermined by senior management, is likely to take several years to complete. It is important to remember that it is really only in the last 10 years that Japanese companies, such as Toyota, have started to reap the benefits of 40 years' investment in TQM.

TQM - here to stay ?

While it is clear that not all companies see TQM as the key to commercial advantage, there are a sufficient number of success stories to demonstrate that, when implemented properly, TQM can have a dramatic impact on a company's profits and market share. This alone should encourage companies to persist in their attempt to achieve total quality, but other factors will also have a bearing. Commercial pressures applied by customers and competitors will force companies to rank quality as top priority. Another fundamental influence will be the growing awareness of environmental performance. Total quality management, by its very nature, requires a company to strive constantly to improve its environmental performance, as an integral part of its quality programme. As indicated earlier, the environment should rank as one of the customers of an organisation, and be treated accordingly.

TQM and the environmental challenge

The pursuit of quality is, of course, largely driven by commercial considerations, but there is an additional moral imperative associated with attaining a high level of environmental performance, and securing the health and safety of employees and members of the public. Many of the features of effective environmental management are indistinguishable from the sound management practices advocated by proponents of TQM.

Organisations of all kinds are increasingly concerned to achieve and demonstrate a high level of environmental performance. The pressure on businesses to "clean up their act" is being exerted from a number of different sources.

Legislative pressure

Environmental legislation is rapidly becoming more stringent as governments and legislators react to environmental catastrophes, such as Bhopal and Chernobyl, and growing scientific evidence that the planet cannot sustain current levels of development unless certain steps are taken to prevent further environmental destruction. Within Europe, tightening of environmental standards is taking place as the European Commission struggles to achieve its twin aims of protecting consumers and establishing a "level playing field" to allow free trade between the Member States of the Community.

Standards of environmental performance are traditionally higher in countries such as Germany, Denmark and Holland. As the completion of the internal market draws nearer, the Commission has sought to use its legislative powers to improve environmental performance throughout the Community and create optimum trading conditions. The result has been the imposition of tougher Community-wide standards for environmental performance to match those already established by certain Member States.

Public pressure

The United States provide an example of the way in which environmental standards can be raised through the implementation of tougher legislation. However, an important aspect of US environmental legislation is that one of the principal driving forces behind its introduction has been pressure from the general public. American concern has no doubt been motivated in part by the number of serious environmental incidents, such as the Love Canal disaster, which have affected the lives of ordinary US citizens. California is a State where high environmental standards prevail, because that is what public opinion has demanded. At least until the United States was hit by recession politicians too have not been slow to realise that "green" policies buy votes.

Pressure groups

Public pressure is also applied through the campaigns of environmental pressure groups, such as Greenpeace and Friends of the Earth. In the United States, certain of these organisations have been particularly active in forcing companies to improve their environmental performance by threatening an unwelcome degree of publicity in regard to what the public might regard as anti-social conduct. In the United States, there is considerably greater public access to information than currently exists in Europe, and in particular, the United Kingdom. US companies have occasionally become victims of their own record-keeping when pressure groups have threatened to "bad-mouth" them in the media, unless the company will agree to improve its environmental performance in accordance with their demands.

Customers and shareholders

Both customers and shareholders are often in a position where they can wield considerable influence over an organisation. As awareness of environmental issues grows, businesses are increasingly experiencing pressure from those they depend on for investment, and those who buy their products. Just as quality improvement is not always the result of self-motivation within a company, improved environmental performance is often a response to external pressure from major customers who are not prepared to have their own reputations tarnished by the poor environmental performance of their suppliers. In much the same way as the drive for quality is forced down the supply chain, the pressure to improve environmental performance is having an impact upon increasing numbers of companies.

Many companies are only too aware of the value of appealing to their customers, and the importance of their environmental performance in this regard. However, the customer, particularly where that means the general public, can often be vulnerable to the onslaught of advertising and marketing literature to which many products expose them. The advertising standards authorities try to curb some of the more extravagant claims which are made, but it is not an easy task. Friends of the Earth, the environmental pressure group, offer their "Green Con Award" on a regular basis to companies who attempt to deceive their public through marketing campaigns and advertising. The European Community is also keen to tackle this issue, although it has adopted a slightly more straight-laced approach than that of Friends of the Earth. The EC proposal is for a system of product labelling, referred to as eco-labelling, which will help consumers distinguish the genuine environmental claims from the more "imaginative". Chapter 4 takes a more detailed look at this scheme and the implications for consumers.

Environmental management systems

All organisational activities, products and services interact with, and have some effect upon, the environment. As businesses become increasingly aware of the need to improve their environmental performance, the use of environmental management systems is becoming more widespread. The components of an environmental management system will normally be inextricably linked with an organisation's corporate management system.

Existing components of a TQM system will be common to operational management, occupational health and safety management, quality management and environmental management. Effective integration and co-ordination of the relevant areas of the corporate management system are essential to ensure consistent decision-making in relation to environmental matters and occupational health and safety. Where there is overlap between the management systems, documentation and records should be shared to avoid duplication and unnecessary bureaucracy, but the inter-relationships should be explained and cross-referenced.

Standards

Chapter 4 looks in detail at the role of BS 7750, the BSI's new standard for environmental management systems, and reviews it in the context of the

Eco-audit Regulation. The British are not alone, however, in producing standards for management systems; there is also the ISO 9000 series of standards from the International Standards Organisation (ISO). In addition, Europe, not to be outdone, has introduced EN 29000 as the European quality standard. Further to the European Community Quality Strategy, the Commission was mandated to establish a common European Standard for the Member States, based upon the ISO 9000 series.

Meanwhile, certain industry associations are developing their own schemes. The Chemical Industries Association (CIA) has introduced a "Responsible Care" programme which commits member companies to improving their environmental performance, in a move towards independent certification of participating companies. Signing up to the guiding principles is now a condition of CIA membership. The CIA has launched two pilot schemes aimed at demonstrating the effectiveness of alternative certification systems: one based upon BS 7750, while the other follows the ISO 9000 series of quality management standards. It appears that the majority of members, many of them international companies, favour the ISO approach.

While the variety of different standards may seem designed to confuse, there are encouraging signs that the ISO 9000 series is likely to reign supreme. Not only is the ISO approach being used increasingly as a basis for the other standards, but it also seems to adopt the most comprehensive approach, particularly for companies hoping to achieve TQM. Unlike the BSI, the ISO decided against drawing up a separate standard for environmental management systems, preferring instead for it to be an adjunct of its quality management standard. Furthermore, the ISO 9000 series addresses the issue of health and safety, as well as environmental management, which BS 7750 fails to do.

Reviews and audits

As part of the general move towards better environmental practice, many organisations have undertaken "reviews" or "audits" to assess their environmental performance. There is currently considerable debate surrounding the definition of the word "audit", and the associated concepts of "review" and "assessment". It is hoped that the Eco-audit Regulation will clarify the situation to a degree, and this whole area of discussion has been expanded in Chapters 1 and 2.

None the less, reviews and audits alone cannot provide an organisation with the assurance that its performance not only meets, but will continue to meet, legislative and policy requirements. An audit or review will simply ascertain a situation at a given point in time; the results cannot guarantee the situation in the future. To be effective, therefore, audits and reviews need to be conducted within a structured management system which addresses all aspects of desired environmental improvement, and can subsequently assist performance measurement.

Environmental improvements in the workplace

Many businesses, particularly in the United Kingdom, are only just starting to wake up to the fact that the excessive demands being made on the environment at large are going to have to stop. However, the issue which more progressive companies are now looking to address is that relating to the environmental quality of the workplace.

A recent report (*The Responsible Workplace*) on business trends, produced by design consultancy DEGW and the Building Research Establishment (BRE), claims that companies are beginning to view environmental performance of premises as part of their competitive advantage. Employees are becoming increasingly aware of the environmental reputation of their employer and the quality of their workplace. The high profile given by the report to environmental issues demonstrates the commercial pressure businesses now face, particularly in light of the growing body of research on employee performance.

Companies seeking to implement a TQM programme, or trying to achieve BS 7750 will need to consider the environmental aspects of their existing premises and buildings they procure in the future – especially if commissioning them. There is a mounting degree of evidence which indicates that offices which are too hot or too cold, have insufficient fresh air and poor lighting, dust, dust mites, airborne bacteria, chemicals from faxes, photocopiers, correction fluids, carpets and cigarettes, can result in a workplace where employees perform 10–30% below par. A working environment of this type can cause employees to suffer lethargy, drowsiness, headaches or upper respiratory tract infections.

Traditionally it has been very difficult to judge the environmental performance of an office, but about two years ago, BRE and the architectural consultants, Energy Conscious Design Partnership (ECD), launched an environmental assessment scheme for new office designs. The new scheme, BREEAM (BRE Environmental Assessment Method), thought to be a world first, is effectively a "green MOT" for buildings. The scheme awards credits for features such as low energy consumption, minimal CFC usage, optimisation of natural ventilation and use of timber from managed forests.

Perhaps one of the most encouraging things about the scheme is the fact that, not only does it help the environment, it actually makes good commercial sense. Experience is starting to show that "green" buildings generally mean reduced running costs. Excessive energy consumption through inefficient heating and ventilation systems and a location away from the public transport network can both contribute to inflated overheads for offices. A "green" office is perceived as a healthy office; it improves productivity and it helps enhance a company's reputation for quality and environmental performance.

Later this year, BRE are introducing a similar scheme for existing offices. It will differ from BREEAM in that it will award credits for the policies, practices and achievements of those operating the building, while taking into consideration the existing equipment and fittings. It is thought that the scheme will appeal to letting agents, owner-occupiers and landlords who refurbish premises between tenants, on the basis of increased marketability, improved working environment and reduced energy costs.

A transnational approach

For some time, individual companies have recognised the importance of environmental management within a TQM system. This view was reinforced in January 1991, when the Global Environmental Management Initiative (GEMI) brought together many of the leading US practitioners of environmental and quality management, from industry, business and government, to participate in "Corporate Quality/Environmental Management: The First Conference". The gathering provided an opportunity for business and industry to consider TQM principles in relation to the complex area of environmental management. Targeted sessions presented pragmatic demonstrations of the way in which the tools designed to apply to TQM can be developed and used for environmental management. GEMI's mission is "To foster environmental excellence by business worldwide", and the conference was designed to provoke an exchange of experience of domestic issues, before moving participants' concerns to the international milieu of the World Industry Conference on Environmental Management, which was held in Rotterdam in April 1991, and for a second time in April 1992.

GEMI has its roots in the idea that environmental quality and total quality are complementary and synergistic concepts. For years, corporate environmental management has featured at fora held by universities and environmental research organisations addressing such questions as how corporations manage the environment and what attributes distinguish the better environmentally-conscious businesses. However, GEMI's aim is to encourage discussions of this nature on the world-wide stage, in order to advance critical thinking and develop tools with which corporations can improve environmental performance, as part of their overall business strategy.

Concluding remarks

While it is apparent that environmental management should be an integral part of TQM, the existence of an environmental management system within an organisation does not necessarily demonstrate achievement of the all-embracing TQM. BS 7750 is not dependent upon prior attainment of BS 5750, and it, therefore, seems that it is technically possible to operate an environmental management system without being committed to a broader quality programme, although it is suggested that BS 5750 provides an excellent basis for achieving the environmental standard.

None the less, there are clearly a number of similarities in terms of the pre-requisites for embarking upon either form of management programme. Commitment at the highest level within an organisation is essential. If management is not absolutely convinced about the changes in company philosophy, it cannot expect the workforce to respond with enthusiasm or motivation. Performance must be measured in order to gauge levels of improvement and the results should be used to encourage employees and inspire them to make even greater efforts in the future.

Regardless of whether the TQM jargon survives, the concepts and philosophy behind the "buzz words" are certainly here to stay. Equally, the environmental challenges which companies currently face will not diminish in size: rather the urgency associated with certain issues is liable to increase. In future, it is increasingly likely that quality management and good environmental performance will be pursued as a common initiative, simply because an increasingly global market will come to expect it; indeed, pressure exerted by competitors, shareholders, customers, legislators, pressure groups and the general public will demand it.

CONTENTS OF CHAPTER 6

The Environment and Workers' Health and Safety

Chapter 6
The Environment and Workers' Health and Safety

Many of the aspects of environmental protection which come to mind first, such as energy saving and re-cycling, seem to have little to do with workers' health and safety. In fact, the interrelationship between the two is such that it would be extremely difficult, if not impossible, to have a proper understanding of either without knowing at least a little of the other.

First, many of the dangers and sources of concern which have prompted the relatively recent flurry of activity in terms of environmental regulation have a parallel impact on working conditions and the safety of workers, which is, of course, the province of the health and safety regime. Take the incidents at Chernobyl and Bhopal, for example. When one reflects on what happened on both occasions, it is easy to regard the main impact of the tragedy as the exposure of the environment to unchecked emissions. Yet in its Annual Report of 1986/87, the UK Health and Safety Commission wrote:

> "Perhaps the most significant *safety* event of the year was one which happened outside our jurisdiction, the Chernobyl disaster. This had immediate and long-lasting physical and psychological effects" (emphasis added).

The short point is that many health and safety and environmental problems have a common derivation.

Even apart from incidents which can be characterised as disasters, the interaction of the two can be seen in relation to, for example, the use of pesticides, and accumulations of dust and airborne contamination which have both health and environmental implications. Indeed, this second example shows that the objectives of regimes aimed primarily at protecting one or the other may conflict: consider the impact on the environment of removing dust and airborne contamination from the workplace by ventilation.

On a practical level, the daily administration of environmental matters within commercial organisations will often be carried out by health and safety officers and representatives since they are the staff most likely to be trained and informed in hazards relating to their workplace. Much of the literature prepared by workers' organisations proceeds on the basis that integration of the two is an important feature of worker participation (see *e.g. Greening the Workplace – a TUC guide to environmental policies and issues at work*, (TUC Publications, 1991) and MSF, *Action on the Environment* publications). In the particular context of assessments and audits (discussed below) there will be an obvious commonality of approach in carrying out investigations and analyses so that a combined approach, even if not essential, offers a clear path to greater efficiency. Consideration of confidentiality may argue against such a combined approach, at least to some extent.

The second reason why no clear line of demarcation can be drawn between health and safety and the environment is that, from a historical point of view, it is fair to regard the current regulatory framework for environmental matters as the young brother of the health and safety regime. This is evident from the fact that the approach currently being adopted towards environmental regulation reflects, in large measure, that which has been reached, after a much longer period of evolution, in the health and safety context.

With this in mind, this chapter looks first at these two points of contact in a little more detail. Then, after a summary of the nature of the legal framework within which health and safety matters are regulated in the United Kingdom, it turns to the first of two areas which may be thought to have most direct relevance to the Eco-audit proposal, namely the use of risk assessments as part of the UK domestic scheme for ensuring that proper health and safety standards are maintained. This necessarily involves a consideration of the impact of the recent wave of EC Directives in the health and safety field which have, in a sense, championed the move away from the setting of specific targets and towards the emphasis on management systems, which is also characteristic of the EC Eco-audit proposal in its latest form (subject to further comment in the Epilogue). This chapter then takes a more detailed look at the extent to which the existing regime provides guidance as to how management should respond to the increasing amount of attention which is being focused on their role in the administration of health and safety; and compares the effect of the EC Eco-audit proposal. Finally, it offers some concluding remarks.

Health and safety and the environment – two sides of the same coin?

The inter-relationship between health and safety and environmental concerns

Reference has already been made to the way in which particular facts can give rise to issues relating both to the health and safety of workers and to the environment. The difference, if there is one, is that the former is concerned more with protecting the personal integrity of workers, whilst the latter deals primarily with protecting the integrity of their physical surroundings.

It might be thought that a further difference lies in the fact that health and safety regimes typically look at risks in the workplace and impose responsibilities primarily on *employers*, as being responsible for the safety of their *employees* by reason of their contracts of employment; whilst the environment tends to apportion responsibility to those directly involved in the creation of the risk concerned. But the reality is that both derive from the same aim: whoever is responsible for the exposure of people or property to a risk should bear responsibility for the consequences of that exposure. This is clearly the case when one recognises that modern health and safety legislation goes beyond the employer/employee relationship and imposes responsibilities on (i) businesses in respect of risks to third parties; (ii) third parties in respect of risks to employees; and (iii) employees themselves. The emphasis is increasingly on the ability to control what is going on and to take action in response.

Any attempt to draw a clear dividing line between the two aspects is bound to fail. Quite apart from the point that most interference with the natural environment arises from business activities, and all business activities involve a workforce of one type or another, an important point of convergence is the way in which they should both influence management decision-making. To expand on this: it has never been the case that society has insisted that personal well-being be protected *at all costs*; and it is unlikely that such a view would ever be taken regarding the integrity of the natural environment. In these circumstances, there is always a balancing exercise to be carried out. The question is how the competing factors – and these include the desire for profitability – are to be weighted. It is difficult to think of any business activity which does not have both a health and safety and an environmental dimension – though, of course, either may be of relatively minor concern – so consideration of both is likely to be unavoidable.

The EC Eco-audit proposal appears to recognise this, as the twelfth recital shows:

> "Whereas the application of environmental protection systems by companies cannot be separated from . . . the need to promote greater protection of the safety and health of workers."

A further recital places particular emphasis on the health implications of business activities:

> "Whereas it is therefore necessary and appropriate to encourage companies to introduce environmental protection systems including audit programmes and to promote systematic, sound information for the public . . . in particular, on the link between health and the environment."

It is surprising, therefore, to find that there is no further express mention of the implications for workers' health or safety of the activities which are to be the subject of review under the audit proposals. This omission is no less strange for the fact that no environmental protection system or audit could sensibly be regarded as complete if it did not look at these aspects. This is particularly true for the occupational *health* aspect of the subject, which includes such matters as pesticides and biological agents, which have a tendency to be overlooked as discussion focuses more on matters of workers' *safety*.

Similarities in the evolution of the two regulatory systems

Legal regulation of workers' safety has existed in a more serious form for a very great deal longer than that of interference with the natural environment. It has also been the vehicle for much of the debate which currently rages as to how far the business of business should be the subject of legal control. It has been said that "the history of occupational safety legislation in Britain is to a great extent, though not exclusively, the history of legislative control over manufacturing industry" (Cmnd 5034, App 5).

This chapter is not the proper place for a detailed consideration of this history, which can be traced back to the Health and Morals of Apprentices Act 1802. But, because the modern approach to environmental regulation and, in particular, the emphasis on management systems which the Eco-audit proposal displays, owes so much to the lessons which have been learned in the health and safety context, it is worth making a few observations on the evolution of regulation in these fields.

The move to legislating for the protection of the environment as well as for the protection of workers is a natural one. Once sufficient mechanisms are in place to protect individuals, including in respect of the indirect effects of damage to the environment on workers, the next step is to protect the environment for its own sake.

The political balancing exercise that underlies the approach to environmental problems is the same as that which has had to be addressed in respect of health and safety legislation. The fundamental question is: to what extent should obligations be imposed on the creators of risk which may have the effect of restricting their ability to engage in the activities which create the risk? The resolution of this difficult issue depends largely upon the "value" which is attached to the "item" at risk. Understandably, life is valued more highly than natural surroundings. So health and safety legislation came first. As society's values change, so does the extent of which aspects of life are worthy of protection.

Logically, the next question is the extent to which restrictions on particular activities are regarded as tolerable. Should hazardous activities be prohibited altogether? Or should obligations be imposed in order to limit hazards or the risks arising from them? Or should financial consequences (penalties, clean-up costs etc) follow (i) the creation of hazards, or (ii) the maintenance of particular levels of risk, or (iii) resulting damage? In the health and safety arena, where these issues have been debated over a very long period, the answer is, as one might expect, a compromise between the three. What limitations on activity are imposed depends upon (i) the nature of the hazard, (ii) the extent of the perceived risk, and (iii) the likely consequences of exposure to the risk.

In the United Kingdom, health and safety legislation has changed its approach to where responsibility for assessing what is or is not a tolerable risk lies. In the early days, decisions were made, to a quite detailed extent, by the legislature. That is to say that a view was formed centrally that (i) particular industries, (ii) particular hazards, or (iii) particular levels of risk were in need of legislative control, and the steps to be taken in particular circumstances were generally prescribed by legislation. Little was left to be decided by individual enterprises which were faced with a range of very specific directions governing the conduct of business activities of their type.

There are a number of fairly obvious criticisms which can be made of this type of piecemeal but, at the same time, specific legislation. Industries which have not been brought to the attention of the legislature (or, indeed, which constitute particularly strong interest groups) may slip through the net altogether whilst, for those business activities which are caught by the regime, the detailed rules imposed may be too harsh or too weak when enterprises are considered on an individual basis. More generally, the effect of such a scheme is likely to be to focus attention simply on compliance with legislation rather than on the overriding desirability of maintaining and improving standards and, perhaps more importantly, the need for effective health and safety management.

To combat these weaknesses, current practice is to insist on formal assessments of hazard and risk, and decision-making concerning necessary preventive and remedial steps being dealt with within each workplace. The role of regulation is, therefore, now seen as being to establish the requirement for assessments, protective measures and so on in general terms by identifying, for instance, types of hazard where risks may exist and where a response may be required.

From a regulatory point of view, the advantage of this is that it enables a greater degree of control to be exercised over individual undertakings; the disadvantage is that employers may not know for certain whether they are doing all that is required of them. From a political point of view, the advantage is that the emphasis is less on achieving particular standards (whether or not in fact they provide adequate protection) and more on achieving an adequate level of protection in each case.

As mentioned above, the consequence of this shift in approach has meant a greater emphasis being placed on the effectiveness of management systems. And this is where the development of the health and safety regime is of particular relevance to the Eco-audit proposal. It seems that regulation of environmental matters has, in a short period, gone through much the same evolution as has health and safety over the last century-and-a-half: there has been a move from specific targets to more generalised objectives; there is an emphasis now on assessments of risk and adoption of control mechanisms on a business by business, and site by site, basis; and there is, therefore, an increased emphasis on proper management systems.

If one is seeking to anticipate what may be down the line in terms of environmental regulation, one major outstanding issue is whether rights of involvement in decision-making, which have been granted by statute to workers in relation to health and safety, will be extended to cover environmental issues at work. In the same vein, questions will arise as to whether workers should be entitled to time off for training, or protection from adverse treatment if they refuse to work when their employer is engaged in environmentally unfriendly activities. In the health and safety field, the position in regard to such matters is relatively clearly established; the solution as far as the environment is concerned has yet to be worked out, and the outcome of the debate is less predictable than that where the very protection of the employees' well-being is at issue.

In this context, it is worth recalling the twelfth recital to the EC proposal (mentioned earlier):

> "Whereas the application of environmental protection systems by companies cannot be separated from the need to ensure appropriate participation by workers in the establishment and implementation of such systems . . .."

It is clear, therefore, that attention has been given to the question of worker participation, which is confirmed by further references to such involvement in Articles 3(1) (commitments of participating companies) and 7(2) (accreditation of environmental auditors) and in Annex I(B) which requires that the framework of environmental protection systems and environmental auditing must cover, amongst other things, "staff information, training and participation on environmental issues".

An outline of the UK health and safety regime

Whilst many, if not most, management and auditing systems will go beyond questions of simple compliance with legislation, all must have an eye towards what the legal requirements are. For this reason, this section contains a brief summary of the UK legal framework for the regulation of health and safety at work.

Overriding principles supported by specific requirements

An appreciation of modern UK health and safety legislation must begin with the general principles laid down by the Health and Safety at Work etc Act 1974 which superseded to a significant extent earlier legislation of limited scope, most notably the Factories Act 1961 and the Offices Shops and Railway Premises Act 1963.

Of these principles, the more significant are:

- Employers must, so far as reasonably practicable, safeguard the health, safety and welfare of employees (s 2).
- Employers with more than five employees must maintain a written statement of safety policy, detailing their general policy and their organisation and arrangements for implementing it (s 2(3)).
- There are prescribed arrangements for consultation with employees and their safety representatives (s 2(4)–(7) (and see the Safety Representatives and Safety Committees Regulations 1977).
- Self-employed workers, employees of other employers and the public generally must not be exposed to dangers or risks to their health and safety as a result of activities carried on by the employer (ss 3 and 4).
- Where certain types of work are being carried out, harmful emissions into the atmosphere must be prevented (s 5).
- Those who design, manufacture or supply products for use at work must ensure, so far as reasonably practicable, that they are safe for use at work (s 6).

– Employees must take reasonable care of their own health and safety, and that of others; and must co-operate with their employer's efforts in regard to health and safety (s.7).

Within the framework of these general principles, there is clearly a need for guidance as to what is required in relation to particular types of business, particular activities and processes, particular substances, particular types of equipment and so on. For this purpose, there exists a wide range of subordinate legislation directed towards specific likely sources of risk. The age and style of the regulations presently in force vary enormously. For industrial premises, much derives from the Factories Acts (even before 1961). More recently, regulations have been made under the 1974 Act itself.

From the point of view of overlap with what may be regarded as mainstream environmental areas, the most important pieces of legislation are probably those in the area of dangerous substances, especially the Control of Substances Hazardous to Health Regulations 1988 (known as COSHH) which came into full effect on 1 January 1990 and have been subject to subsequent amendment. These regulations are of entirely general application, and cover all workplaces and apply to all substances *hazardous* to health – which means only that they have the *potential* to cause harm, not that such harm is likely or serious – in the form in which they are used in the workplace with the exception of lead, asbestos, radioactive substances and substances in underground mines (all of which are covered by other legislation).

The principal requirements of the COSHH regulations are that employers should: (i) assess the risks arising from the use of the substance, (ii) introduce measures to prevent or control the risk, (iii) ensure that those measures are implemented, (iv) in some cases introduce health monitoring and surveillance, and (v) provide appropriate training and information to workers.

Also of importance are the amended Control of Industrial Major Accident Hazards Regulations 1984 (known as CIMAH) which derive from the Seveso Directive. These apply to activities which are regarded as representing a particular hazard to the workforce or the neighbourhood by reason of the processes being carried out, the use being made of toxic substances or the storage of dangerous substances. Typical examples are chemical and petrochemical plants – nuclear installations are subject to separate legislation. Through these regulations, particularly stringent demands are imposed for, first, the identification of hazards and, secondly, the demonstration, not only that (i) steps are being taken to prevent hazards and limit their consequences to persons and the environment, but also, (ii) where the hazard is regarded as being particularly significant, that procedures are in place for dealing with major accidents both on-site and off-site (though the latter is primarily the responsibility of the relevant local authority).

Since the 1974 Act, the Health and Safety Commission (HSC), which is largely responsible for developments in this area of the law, has been undertaking a continuing process of modernisation which has been greatly influenced by (and, indeed, which has itself influenced) developments at EC level. The COSHH regulations are just one early example of legislation to implement an EC Directive.

Imminent changes

During the European Year of Safety, Hygiene and Health Protection at Work (March 1992 to February 1993) the influence of EC developments has been particularly pronounced since no fewer than six Directives of more or less general application were due for implementation by Member States by 1 January 1993, together with a number of others. The pace of legislation at the Community level has been given a fresh impetus by the inclusion of Article 118A, which provides for the adoption of Directives through a form of majority voting, in the Treaty of Rome. So these six Directives represent only the start of the new initiative: others are following dealing with construction and engineering works, work in the extractive industries and – significantly in the context of a discussion of the relationship between health, safety and the environment – biological agents, to name but a few.

The six major Directives referred to above follow a broadly similar structure to that of the post-1974 UK legislation. That is to say there is a Framework Directive setting out overriding principles and matters of general importance, supported by five Directives addressing identified aspects of the protection of workers' health and safety (the Workplace, Work Equipment, Personal Protective Equipment, the Manual Handling of Loads and Display Screens). All will be implemented by regulations made under the 1974 Act since the Health and Safety Commission has taken the view that no fundamental revision of the UK legislative regime is required. (Details of the reasoning behind the proposals for implementation can be found in HSC Consultative Documents CD34, CD43, CD35, CD36, CD38 and CD 42 respectively.)

Nevertheless, the changes which the Directives make are significant. It has been said that the effect of the Directives is to make explicit what is implicit in existing legislation and, as such, their requirements ought to reflect existing good practice. However, the emphasis which is placed on the need for assessments of risk on a case by case basis, and the duties to consult with and inform employees about the risks to which they are exposed – and measures being taken for their protection – is likely to have a major impact on the way in which businesses organise management of health and safety matters.

Enforcement and penalties

Enforcement powers for health and safety legislation are conferred generally on the Health and Safety Executive (HSE), which is the operational arm of the HSC. By way of exception, local authorities are the enforcing authority for most retail, office, hotel and recreational operations. In addition, the HSE has, by agency agreements, transferred enforcement powers in relation to certain types of undertaking where specialist activities are carried on, such as in relation to the nuclear industry, the use of pesticides and pollution.

So far as means of enforcement are concerned, the enforcing authorities have wide powers, including entry, inspection and questioning, and can issue improvement notices (requiring steps to be taken within a given time period) or, where the inspector believes there is a serious risk of personal injury, prohibition notices can be issued (stopping work until specified measures have been implemented).

Prosecutions have traditionally been regarded as a last resort approach to securing compliance, though there have been recent signs that this attitude is changing: note the tone of the following statement, taken from the HSE Director General's foreword to the 1989/90 Annual Report of the HSC and HSE:

> "We welcome, too, signs that the courts are now prepared to act severely in cases of serious dereliction of health and safety responsibilities. Not only did the average level of fines imposed by magistrates courts increase, but there were some very large fines imposed in particular cases in the Crown and Scottish courts."

The amount of fines is one area where it seems that health and safety regulation has been substantially influenced by the approach to environmental matters. For offences dealt with by magistrates courts – and hitherto most of them have been – the maximum fine has been £2,000, though statistics show that average fines have been very much lower. Pressure for greater emphasis to be placed on achieving higher health and safety standards, and a desire for parity with environmental offences, has meant that, during 1992, maximum fines, even before magistrates, have risen to £5,000, with certain offences attracting a maximum fine of £20,000 (see the Criminal Justice Act 1991, s 17(1) and s 4 of the Offshore Safety Act 1992, which does not deal only with safety in the offshore industry).

Fines in the Crown Court remain unlimited and can be quite substantial in practice. So, from a financial point of view, given the obstacles to insuring in respect of liability for criminal penalties, a failure to comply with health and safety legislation must now be regarded as being just as great a financial risk as it has become in respect of environmental legislation.

The risk, however, extends much further. Although to-date the courts have imposed few custodial sentences on individuals who have been convicted on a personal basis, the likelihood must be that this option will be pursued more frequently as attention focuses to a greater extent on directors, managers and others with responsibility for organising workplace compliance. The HSE has made it clear that they intend to take advantage of section 37 of the 1974 Act, which allows such personal prosecutions for directors, managers and similar officers of a company, whenever they can connect defaults by senior management with non-compliance with statutory requirements.

On a similar note, June 1992 saw the first use by the UK courts of the Company Directors Disqualification Act 1986 to disqualify a director (in this case for two years) for inadequate management of health and safety. Culpable health and safety mismanagement is, therefore, on the same plane as culpable financial mismanagement. There is no reason to believe that culpable environmental mismanagement will be treated any differently. As David Eves, HSE Deputy Director General, said:

"It is clear that the courts are beginning to share HSE's views of the importance of health and safety management. Directors and managers should take note. This case demonstrates that disqualification of directors is a sanction that the courts will use in appropriate circumstances.".

Assessments in the health and safety context

In Chapters 1 and 2, consideration is given to some of the key elements of any environmental audit, elements which seem to be present in one form or another in most of the serious proposals for such auditing. It is clear that they go beyond mere considerations of compliance with statutory requirements, though they will need to include an appraisal of the risk and extent of non-conformity.

The Eco-audit proposal, of course, covers questions as to the adequacy of the environmental management system in place and, to that extent, it is perhaps mis-named. It seems right, therefore, to regard an environmental audit as at least including something in the nature of a risk assessment exercise, though different forms of audit may be directed towards the identification and assessment of different types of risk. The risk of prosecution may be uppermost in business' minds, but if "greenness" is an objective which is regarded as valuable for reasons of public profile, marketing or, perhaps, conscience, a more general appraisal of the risk of interference with the natural environment will be required. Here we consider this "assessment" aspect; below we give further attention to the question of management systems and audits of them.

Any enterprise will find itself facing difficult questions of balance. Reference has already been made to some of the factors which have to be weighed against each other. What hazards are acceptable? What levels of risk are acceptable? (Indeed, what does "acceptable" mean, in this context?) How can respect for the natural environment be reconciled with the understandable desire for profitability?

It is to such questions that Article 3(1) of the EC proposal is directed when it speaks of the need for participating organisations:

"to examine, at the highest management level, all the findings of the audit and, in the light thereof and within the framework of their environmental programme, set objectives and measures aimed at continuous improvement of their environmental performance."

And the environmental statement which the proposal requires must include

"an evaluation of the environmental performance of the protection system implemented at the site considered."

So the assessment concept represents a point of convergence for the two areas – though the voluntary character of the EC Eco-audit assessment represents an important difference from the mandatory requirements for health and safety – because the concept of risk assessment lies at the very heart of modern health and safety law. Indeed, there is an increasing tendency to talk not of "health and safety law" but rather of "risk management": the consequent emphasis on management systems is considered further below. Even beyond what legislation explicitly requires, risk assessments are clearly an indispensable requirement of good practice. This is borne out by the fact that the HSE has described the principles set out in its recent publication *Successful Health and Safety Management* (1991), which is largely directed towards risk identification and control, by saying that:

> "The advice given here will be increasingly used by HSE inspectors as a basis for testing the performance of organisations against the general duties of the Health and Safety at Work etc Act 1974."

The most well known example of legislation explicitly requiring an assessment is the COSHH regulations discussed above. But there are others – see, for example, the Control of Asbestos at Work Regulations 1987 and the Noise at Work Regulations 1989; and Directives resulting from the present EC initiative focus very much on assessments within the workplace in respect of particular hazards, as has been discussed.

Statutory requirements for assessments

The HSC has said (CD34) that:

> "The . . . requirement to assess risks for the purpose of identifying protective and preventive measures is . . . implicit in section 2 of the HSW Act and explicit in many other regulations such as the Control of Substances Hazardous to Health (COSHH), Noise, Lead and Asbestos."

That the requirement is explicit in the regulations referred to is undoubtedly true. That it is implicit in section 2 of the Act is more dubious. Here is not the place to debate the proper interpretation of the section, but what can be said with a fair degree of confidence is that, whatever the strict legal position, few employers will have regarded the general words of section 2 (described above) as necessitating any formal assessment exercise.

It seems that that position will now change following the regulations to implement the Framework Directive (89/391/EEC): regulation 3(1) of the Management of Health and Safety at Work Regulations 1992 provides that:

> "Every employer shall adequately assess –
> (a) the risks to the health and safety of his employees to which they are exposed whilst they are at work; and
> (b) the risks to the health and safety of persons not in his employment arising out of or in connection with the conduct by him of his undertaking
>
> for the purpose of identifying the measures he needs to take to comply with his duties under the relevant statutory provisions."

So, if business has escaped hitherto the burden of a full-scale consideration of the impact of its activities on the health and safety of its workers and other third parties, it seems unlikely that it will be able to do so for much longer. And given the breadth of regulation 3(1)(b), and the reality that most impacts on the environment also affect people, at least indirectly, the result may be that a statutory requirement for at least limited environmental assessments will have been introduced by a sidewind.

What does an assessment involve?

Feedback from employers indicates that, in some quarters at least, the requirement for assessments is not well understood and that this leads often to one of two different responses: there are those who go through the formalities, particularly in terms of identifying hazards, without really knowing why they are doing so or what conclusions they should draw; and there are those who abdicate the whole exercise to external consultants (and then criticise the regulations for imposing an excessive financial burden).

Given that safeguarding the health and safety of individuals can be achieved at a range of different levels, and that few legal requirements are absolute in the sense that they must be taken irrespective of financial implications, it seems that the most important qualifications for those involved in carrying out the necessary assessments are common sense and experience. There may well be occasions when some form of numerical quantitative analysis can assist, but it is difficult to see how that can offer more than a partial solution.

It is understandable why there is so little official guidance as to precisely what is required in carrying out an assessment: though one may doubt the wisdom of the different terminology used to describe the nature of the assessment which is sometimes described as being required to be "suitable and sufficient" (see the COSHH Regulations) and at other times "adequate" (as for asbestos and noise). Regulations and guidance notes can specify objectives, and can give clues as to relevant considerations, but it is in the nature of the more flexible approach currently being pursued that rules cannot prescribe a course of action appropriate to every case.

Indeed, the nature of the assessment required by legislation often varies depending upon the particular context: under the regulations implementing the Personal Protective Equipment (PPE) Directive, assessments are required to ensure that PPE is in fact suitable for its task; whereas the assessment under the regulations for the Display Screens Directive is intended to address risks to users of workstations; and, for the Manual Handling of Loads, assessments come into play only if an employer is satisfied that he cannot readily change the method of working so as to eliminate the risk of injury. Implementation of the Framework Directive will require a generalised form of risk assessment, in effect a safety audit – see the Management of Health and Safety at Work Regulations 1992.

The simple fact is that where mandatory measures exist, one can ascertain whether or not they are being achieved; but beyond that, it is difficult to say what is or is not the proper way to deal with a particular hazard. It is, ultimately, a matter of judgement. Therein lies the difficulty with which the health and safety world continues to grapple: how can any employer be certain that he will not be prosecuted for failing to take proper steps to safeguard his workforce, when the adequacy of the steps which he has taken are only likely to be judged retrospectively when an incident has occurred? As a matter of principle, the mere fact that an incident occurs should not be taken to mean that he has failed in his responsibilities. Equally, it must be clear that the mere fact of having carried out an assessment will not protect him from liability in every case.

For its part, the UK Government has been critical of those who spend too much money on engaging external consultants to do what it regards (perhaps, for political reasons) as largely a common sense exercise; but at the same time it has tried to make the exercise more straightforward by including in legislative proposals what can be regarded as default standards *i.e.* there are sometimes quite specific statements as to what is considered the norm in terms of preventive and protective measures. A good example of this is the Guidance (L23) provided with the Manual Handling Operations Regulations 1992. It remains to be seen whether this will be counter-productive: it will if the result of this form of assistance is that employers take the easy course and make the assumption that their situation is within "the norm" and simply adopt the guide-line standards without giving the matter any further consideration.

Perhaps the most important lesson to learn from this state of affairs is that formalised risk assessment is still at a relatively early stage of development, and that the adoption of management techniques (such as audits) which it involves is bound to result in a degree of uncertainty. That will be all the more the case where the audit is not for compliance with objective standards but rather involves comparisons with individual perceptions as to what is or is not acceptable from a public relations or a moral standpoint, which is often the case where interference with the environment is concerned. What one can say is that the cost/benefit analysis which risk assessment of this kind involves ought, at least in principle, to be easier to operate in the environmental field because, in many cases, the harm that is sought to be avoided is the need for expenditure on clean-up costs. That is far more readily quantifiable than the value of human life, the preservation of which is the ultimate goal of health and safety good practice.

All this is not to say that risk assessment techniques should not be used. Rather, it is important that their character be properly understood. In particular, it is fundamental that their dependence on subjective criteria be recognised and that any consideration of the conclusions of such assessments includes an analysis of the assumptions on which they are based.

Before leaving the subject of assessments, mention should be made of the concept of "responsible care", the implementation of which, in UK terms, has been led by the Chemical Industries Association (CIA). The philosophy underlying the responsible care programme is a commitment on the part of members of the CIA not only to meeting statutory requirements in the field of health, safety and the environment but, beyond that, to assess the impacts of their activities in those areas and to work to acceptably high levels of achievement – what is "acceptable" being determined through consultation with the public and relevant authorities – and to operate according to industry best practice. In this way, the programme's main objective is to secure the full integration of safety, health and environment issues in business planning. The close relationship with notions of Total Quality Management (TQM) will be evident.

The emphasis on management systems

The effectiveness of management systems is clearly a further key element of the EC Eco-audit proposal. Indeed, the proposal has been criticised on the basis that management systems will constrain/influence management in its decision-making: they will, of course, because that is precisely their aim. This parallels the trend in the regulation of workers' health and safety. The short point is that experience shows that however many specific requirements are imposed, and however much legislation is in place, there will always be a limit as to what can be achieved in terms of reducing the frequency and severity of accidents without effective management of health and safety.

This in turn requires acceptance by senior management that health and safety is worthy of management time. The tactic which has, therefore, been employed in the United Kingdom is to make every effort to ensure that health and safety is taken seriously by senior management by, for example, increasing the level of fines to a point where they can have a real impact on profitability, the more frequent use of custodial sentences for directors and, more recently, through the possibility of disqualifying directors. The same objective also underlies the shift of emphasis towards assessments of the kind described above.

Certainly in the United Kingdom the enforcing authorities' current approach is to seek to drive home to businesses the point that responsibility for reducing accidents lies with the business rather than with the authorities. And whether or not one believes that a safe organisation is a profitable one, it is undoubtedly true that the costs of accidents are high.

Assessing the adequacy of management techniques

Prior to implementation of the health and safety framework Directive, there was no general mandatory UK standard for health and safety management. Whilst it was argued, as in relation to the narrower question of risk assessments, that a requirement for sound management is implicit in the general principles of the 1974 Act, it is difficult to believe that such principles could have been used as a firm basis for taking action against organisations where there is less than adequate management.

Probably the closest that the legal regime came to this is in the requirement for a written safety policy document which is to be found in section 2(3) of that Act. It provides:

> "Except in such cases as may be prescribed, it shall be the duty of every employer to prepare and, as often as may be appropriate, revise a written statement of his general policy with respect to the health and safety at work of his employees and the organisation and arrangements for the time being in force for carrying out that policy, and to bring the statement and any revision of it to the notice of all of his employees."

But what exactly must a safety policy document specify and, in particular, what "organisation and arrangements" will be regarded as meeting the requirements of the statute? Legislation does not answer these questions, so it has been left to HSE guidance publications to fill out more precisely what is required. What do they say? Reference is typically made to two leaflets entitled *Writing a Policy Safety Statement* and *Guidance on the Implementation of Safety Policies* which give a fairly clear indication as to what is meant by the various terms which appear in section 2(3) and the range of matters which should be taken into consideration. What the HSE's leaflets do not seek to do is to influence the adoption of any particular overall management approach.

Broadly speaking, the requirement for a written safety policy (and its contents) mirrors the EC proposal's statement that environmental protection systems:

> "will include, in particular, the following elements, formally stated in writing: an environmental policy, environmental objectives, an environmental programme and an environmental management system . . .."

A possible difference lies in the extent to which such matters must be publicised. Whilst the EC proposal requires much to be made available to the public through the environmental statement, it seems that there is no need to explain the management system in any document for general consumption. In the safety context, on the other hand, the management system most definitely is to be dealt with in the written policy document, but publicity is limited to bringing the statement to the attention of employees.

The HSE's more recent publication, *Successful Health and Safety Management* (1991), goes much further and offers clear guidance on effective health and safety policies, organising and planning for health and safety and measuring, auditing and reviewing performance which, together, it describes as "the key elements of successful health and safety management".

The Management of Health and Safety at Work Regulations 1992 seek to remedy the deficiency in the law and, as well as explicitly requiring a general assessment of risks to health and safety, as described above, impose a number of specific requirements including:

> "4. (1) Every employer shall make and give effect to such arrangements as are appropriate, having regard to the nature of his activities and the size of his undertaking for the effective planning, organisation control, monitoring and review of the preventive and protective measures
>
> (2) Where the employer employs five or more employees, he shall record the arrangements referred to in paragraph (1).
>
> . . .

6. (1) Every employer shall, subject to paragraphs (6) and (7), appoint one or more competent persons to assist him in undertaking the measures he needs to take to comply with the requirements and prohibitions imposed upon him by or under the relevant statutory provisions."

Additional moves in this direction can be seen in the Offshore Safety Act 1992 – which extends the application of the provisions of the 1974 Act to cover offshore installations – and its associated regulation: the Offshore Installations (Safety Case) Regulations 1992. The Guidance Notes to these regulations are particularly instructive in understanding what will or will not be regarded as an acceptable management system.

Finally, mention must be made of the Temporary or Mobile Construction Sites Directive which contains perhaps the most clear description of how a safety management system should be arranged, by identifying particular posts with specified responsibilities. The draft regulations in the United Kingdom, set out in Consultation Document CD51, show the tension between such a rigid approach and the more flexible attitude traditionally adopted in the United Kingdom. The trends are clear!

Whilst the wider publicity requirements and the more extensive subject matter of the Eco-audit environmental statement mean that it would be a mistake to draw too many analogies between it and the statutory requirement for a safety policy statement, it is worth noting recent proposed legislative changes in the United Kingdom. These changes would represent a move towards the wider requirements of Article 5 in relation to both health and safety and the environment, remembering that the environmental statement must be prepared as the basis of an environmental audit, one of the objectives of which is to assess observance of existing regulatory requirements.

Under the Corporate Safety and Environmental Information Bill proposed by Jeff Rooker MP in December 1991, the statutory directors' report would have to include details of convictions under health and safety and environmental protection legislation, together with details of notifications under the Reporting of Injuries, Diseases and Dangerous Occurrences Regulations 1985, and compensation payments made.

This Bill had widespread cross-party support prior to the General Election of 1992. Its aims have been given a new lease of life through the "Right to Know Bill", sponsored by Mark Fisher MP. Amongst other provisions, the Bill would oblige companies with 50 or more employees to provide, in their annual reports, details of the following:

- the numbers of occasions on which enforcement action was taken against the company under a range of legislation including in relation to public safety, health and safety at work and environmental protection legislation;
- the number of deaths and serious injuries to employees during the year, expressed as injuries per 100,000 employees; and
- the total amount of compensation paid in respect of injuries resulting from the company's activities and products, and details of legal costs.

There is, of course, a world of difference between even clear statements of management structures and objectives and what is actually done. Hence the shift, in a slightly different context from talk of Quality Assurance (BS 5750 etc) to discussion of TQM, which places greater emphasis on what actually happens on a day-to-day basis (see Chapter 5). Effective management is not just about systems which exist in principle. It is about achieving results through the effective operation of such systems. Most auditing systems are aimed, at least in part, at reviewing performance and the EC Eco-audit is no exception. One example of how the pursuit of a TQM approach can result in a dramatic improvement in levels of health and safety achievement is the progress made by Stocksbridge Engineering Steels over a period of some five years. How it succeeded in, at the same time, increasing productivity by 26% and reducing its lost time accidents by 55% is fully described in Health and Safety Information Bulletin 202 (October 1992).

By way of comparison with the Eco-audit proposals, and indeed BS 7750, it is worth paying particular attention to the section of *Successful Health and Safety Management* (referred to above) which deals with auditing and reviewing performance, since that is probably the closest to an authoritative guide found in the health and safety context. It is particularly important to note that this is seen as being far wider than an assessment of "mere" compliance with specific statutory provisions. On the question of what characteristics an effective auditing system should display, the following are identified:

(i) They should be undertaken by competent staff, independent of the area being audited.

(ii) They are designed to assess:

   (a) the intent, scope and adequacy of the health and safety policy (and this means the actual policy and not just whatever document it is set down in);

   (b) the organisation itself, which includes: whether responsibility for health and safety is accepted by line managers, whether arrangements are in place to inform and involve all employees in respect of the efforts being made, and whether steps are taken to ensure employees' competence;

   (c) planning and policy implementation both generally and in respect of the particular targets and standards which are set: are the standards actually being achieved? and

   (d) the adequacy and relevance of the systems for measuring performance, reviewing procedures and taking account of lessons learned.

The similarities between this and the scope of an environmental audit to be found in the EC proposal are self-evident and provide further support for the view that, from the standpoint of efficiency, there is much to be said for adopting more or less similar management systems (in respect of auditing more widely) in respect of health and safety, and the environment. On the question of what management approach is appropriate, the principles described in *Successful Health and Safety Management* have much in common with BS 7750 (discussed further in Chapter 4), and it may very well be that companies are able to use BS 7750 (or its international equivalent) as a means of integrating the management of health, safety and the environment. Certainly, any attempt at TQM will require both of these areas to be dealt with in a co-ordinated manner.

Finally, the provisions of the Corporate Accountability Bill, now a couple of years old but a good indicator as to present thinking in this area, include requirements for the appointment of independent safety auditors whose duty it is to report to the company in its general meeting "on the policies, practices and record of the company with regard to the health, safety and welfare of persons (whether or not persons in the employment or service of the company)". The report would have to state "whether in the auditors' opinion the company has complied in all material respects with best available practice in matters relating to the health, safety and welfare of persons" and would have to give details of respects in which there had not been such compliance. The breadth of these proposals could encompass the whole of the management of health and safety. That is precisely their intention: culpable health and safety mismanagement must be treated as being as significant as culpable financial mismanagement.

Concluding remarks

It will by now be clear that the effective management of health and safety on the one hand and of environmental protection on the other cannot be separated if any serious attempt is to be made at either. Quite simply, they have too much in common. Not only do many activities create potential risks both for individuals and to the environment, but the understanding of the way in which any business functions, which is required for the management of each, is in many respects the same. So a co-ordinated policy is suggested both by the reality of business life, and by arguments for efficiency.

One word of warning: it should not be assumed that in all cases a combined health and safety and environmental management structure is appropriate. Considerations of confidentiality are just one reason why structural efficiency should not be regarded as the only goal.

Furthermore, the parallel, though at times uneven, development of the legal regulation of both areas means that they have much to teach each other. Public pressure for change has been motivated by a concern to ensure that the value of human life and protection of the natural environment are not entirely displaced by the desire for corporate profitability. The response of the legislature has been to make the consequences of overstepping the mark more of an impediment to corporate success.

Again, similarities can be found in the detail of the regulatory approach. There is an increasing tendency to focus on deficient management which can be seen clearly in the approach being adopted to enforcement and in the structure of the modern regulatory framework. In the health and safety arena, there is a move towards the need for individual employers to take decisions as to what steps should be taken in order adequately to protect the health, safety and welfare of their employees through the vehicle of risk assessments. Similarly, the adoption of evaluative techniques represents a key feature of the EC Eco-audit proposals. Conversely, there have been suggestions during recent discussions as to the scope of the Fourth Community Action Programme on Health and Safety that something akin to the Eco-audit Regulation should be established for health and safety, in order to achieve a coherent approach to working environment issues.

More generally, the stance adopted by the HSE in the United Kingdom shows that management techniques will be under close scrutiny in the new era, which is precisely the aim of the EC proposal. The accumulated experience and learning in respect of each offers a fund of knowledge from which both can draw.

It may be true that the Eco-audit proposals will be a matter for voluntary participation and, in that sense, can be distinguished from what health and safety legislation requires, at least in the United Kingdom. But, from a practical point of view, that distinction is more apparent than real. Just as satisfying the requirements of BS 5750 (and its international equivalent) has, in many industries, become an essential prerequisite to the ability to compete effectively in the market, it is not unlikely that the same will become true of the Eco-audit. There are already clear signs that demonstration of a suitable approach to health and safety management is being regarded, in the construction industry, as a pre-requisite to inclusion on tender lists. And to argue, from the voluntary character of the Eco-audit requirements, that good business practice does not require the effective and co-ordinated management of health and safety and the environment, is to live in a world far removed from reality.

CONTENTS OF CHAPTER 7

Environmental Assessment and the Problems of Contaminated Land

Chapter 7
Environmental Assessment and the Problems of Contaminated land

Environmental assessment, sometimes called Environmental Impact Assessment (EIA), is another important tool for managing environmental issues in the context of land development and use. Its discussion is valuable both in the context of proper environmental management and environmental openness. Whereas the Eco-audit is designed to assess the *actual* effect of existing uses and processes upon the environment, Environmental Assessment is used to assess the *potential* effects of a proposed development on the environment. In simple terms, environmental assessment looks at prevention at source, whereas the Eco-audit looks at compliance and damage limitation. In so far as the standards of published environmental statements have been used as a measurement of quality of the work of intending environmental auditors, the relevance and interaction are self-evident.

Environmental assessment has been introduced in the United Kingdom thanks to Directive 85/337 EEC of 27 June 1985 on the assessment of the effects of certain public and private projects on the environment (in this chapter referred to as "the Environmental Assessment Directive"). The original idea came from the United States where the National Environment Policy Act 1969 (NEPA) provides that where a federal agency proposes a development it should prepare an environmental impact statement for the purpose of considering its possible impacts on the environment before a decision is taken whether to proceed with the development.

When the Environmental Assessment Directive was first conceived it followed NEPA in certain ways, for example the onus was on the developer to produce a full EIA. In other ways, initial proposals for the Environmental Assessment Directive exceeded NEPA, for example by requiring environmental assessment for *all* projects which would be likely to have significant environmental effects.

The proposed Environmental Assessment Directive did not have an easy passage. Some 20 drafts were prepared by the Commission between 1977 and the formal proposal in 1980. The Directive was further watered down in passage so that in its final form Environmental Assessment does not apply to all development. Rather it applies only to specified categories likely to have significant environmental effects. The developer is not required to carry out his own EIA, but rather to provide the specified information in the form of an environmental statement, to enable the decision maker to carry out an EIA.

The Directive was notified to the Member States on 3 July 1985 who were required to implement its provisions within three years.

What is environmental assessment?

Environmental assessment is much more than the production by the developer of the environmental statement containing the specified information. Environmental assessment is a process whereby the decision-maker who is responsible for giving consent to a proposed development reviews its likely effects upon the environment. There are a number of stages including:

(i) The developer or, sometimes, the decision-maker, deciding whether an environmental assessment is required.
(ii) The developer establishing the scope of the environmental assessment and thus the contents of the environmental statement.
(iii) The developer approaching any relevant organisations who may have information which would assist in the preparation of the environmental statement.
(iv) The developer preparing the environmental statement containing the specified information.
(v) The decision-maker considering whether any further information is required and seeking this from the developer.
(vi) The decision-maker consulting with the public and relevant organisations.
(vii) The decision-maker evaluating *all* information received and representations made to assist with the decision as to whether to grant consent for the proposal.

It is apparent that the environmental statement is only a part of the whole process, albeit a very important part. The aim of the Directive is to ensure that proposed projects in all EC Member States which are likely to have a major effect on the environment are carefully scrutinised and consent is only granted once the environmental effects have been properly assessed. The Directive aims to create similar national laws in the field of environmental assessment to prevent unfavourable competitive conditions and there are provisions to harmonise the implementation of the Directive throughout the Member States. The philosophy behind the Directive is "that the best environmental policy consists in preventing the creation of pollution or nuisances at source rather than subsequently trying to counteract their effects", coupled with "the need to take effects on the environment into account at the earliest possible stage in all technical planning and decision making processes." Businesses which intend to embark on development projects should be aware of the ramifications of the Directive in order to enable proper resources to be taken into account at the right stage, and thereby to avoid costly mistakes.

The Environmental Assessment Directive

The Environmental Assessment Directive sets out a basic framework for environmental assessment to be implemented in each Member State. It lists certain categories of development where environmental assessment is required, although Member States are not restricted to these categories. The classification is found in two annexes. For those projects set out in Annex I to the Directive, environmental assessment is mandatory in all cases but, in respect of those projects set out in Annex II, Member States are given a discretion and environmental assessment is only to be required for projects where Member States consider "that their characteristics so require".

The nine categories of project for mandatory assessment include construction of such facilities as crude oil refineries and thermal power stations, and motorways and certain other roads over a certain length. It is generally very easy to decide whether or not environmental assessment is required here; one merely looks at the type and size of development proposed.

For projects where assessment *may* be required, it is necessary to look at the 12 categories of project which are set out together with a detailed list of projects within each category. The 12 categories include agricultural projects, certain infrastructure projects and projects proposed by the extractive or energy industries.

Where environmental assessment is required a developer must provide at least the specified information which is:

(i) A description of the project comprising information on the site, design and size of the project.
(ii) A description of the measures envisaged to avoid, reduce and, if possible, remedy significant adverse effects.
(iii) The data required to identify and assess the main effects which the project is likely to have on the environment.
(iv) A non-technical summary of the above.

The Directive provides for consultation with interested "authorities" and the public and provides that relevant authorities should make information in their possession available to the developer. All of the information gathered must be considered by the relevant authority in making its decision on whether to grant consent. An EIA must consider the direct and indirect effects of a project upon:

(i) Human beings, fauna and flora.
(ii) Soil, water, air, climate and the landscape.
(iii) The inter-action between (i) and (ii) above.
(iv) Materials, assets and the cultural heritage.

This is indeed a wide brief!

The Directive does not specify the physical form of the EIA. In particular, there does not appear to be a requirement for a written EIA to be produced, and the Directive only requires reasons for a decision to grant or refuse consent to be given "where the Member States legislation so provides."

The Directive required the Commission to send a report on the implementation and effectiveness of the Directive five years after its notification (*i.e.* 3 July 1990). At the time of writing this report was still awaited.

Implementation of the Environmental Assessment Directive in the United Kingdom

One of the reasons that the Directive took so long to move from the first draft proposal in 1977 to its final form in 1985 was a number of objections including opposition from the UK Government who were of the view that a formal Environmental Assessment procedure was unnecessary in the United Kingdom. It was the government's contention that environmental matters were considered in any event in the development consent procedures under the Town and Country Planning Acts. In addition they had a number of specific reservations including one which has, to a certain extent, proved to be justified, namely the difficulty of defining which projects should be subject to assessment.

The proposed Directive was substantially amended to take account of a number of the Government's objections, and in November 1983, due in a large part to the support for the proposal in the House of Lords, the government withdrew its opposition. The Directive has been implemented in the United Kingdom, at least in part by tacking on regulations to the existing planning procedure. William Waldegrave, the relevant minister at the Department of the Environment (DoE) at the time when implementation was considered, stated that the government wished to "ensure that the requirements of the Directive are met without placing unnecessary burdens on the developer or unnecessarily complicating the planning process."

Accordingly, rather than passing primary legislation, the Directive was implemented by the passing of regulations under the European Communities Act 1972, the principal regulations being the Town and Country Planning (Assessment of Environmental Effects) Regulations 1988 (referred to in this chapter as "the Planning Regulations"). The Planning Regulations came into force on 15 July 1988 (only a few days after 3 July 1988, the required three-year deadline for implementation set out in the Directive) and apply to applications for planning permission to the Local Planning Authority, applications referred to the Secretary of State and appeals to the Secretary of State where the relevant *application* was received by the Local Planning Authority on or after 15 July 1988. This provision has caused some considerable controversy as will be seen below. For those projects which do not require planning permission under the Town and Country Planning Acts it has been necessary to introduce a series of 19 other regulations which deal with environmental assessment.

Environmental assessment as required by the Planning Regulations

The majority (approximately 60–70%) of all environmental assessments carried out pursuant to the Environmental Assessment Directive in the United

Kingdom are carried out under the Planning Regulations. The Planning Regulations follow, to a large extent, the provisions of the Directive. There are two types of projects: those requiring mandatory assessment and those where assessment may be required if the project is likely to have "significant environmental effects". The developer must provide an environmental statement containing specified information, various bodies or statutory consultees and the public must be consulted and be given an opportunity to inspect the environmental statement and to comment. All of the information gathered must be taken into account by the local planning authority, the Inspector, or the Secretary of State in forming their judgment on whether the development should go ahead.

The decision as to whether or not permission is to be granted remains to be taken under the legislation relating to planning. Therefore, environmental considerations are not the sole determinant; other material planning considerations will be important. This means that, at least in theory, it is open to a planning authority to grant permission for development which will have a significantly adverse environmental effect if there are other material considerations to which they give a higher priority.

The detailed procedural requirements are set out in the Planning Regulations and in DoE guidance. In summary, where the Planning Regulations apply and environmental assessment is required, the planning application should be submitted in the usual way but it must be accompanied by an environmental statement prepared by the developer and the usual eight-week period for determination is extended to 16 weeks. Applications subject to environmental assessment must also comply with the planning publicity requirements (*i.e.* press and site notices will be required). As well as making copies of the environmental statement available to members of the public, the developer must provide sufficient copies for all of the statutory consultees (see below). The environmental statement must be served on the statutory consultees, and planning permission cannot be granted until the environmental information has been taken into consideration. Where the Local Planning Authority believe that there is insufficient information provided in the environmental statement they can require provision of further information or evidence in verification, but at the time of writing there is no requirement for this further information to be distributed to the statutory consultees.

One of the most important steps in the process is what is known as "scoping" the environmental statement. If the scope of the assessment (*i.e.* the issues relevant to it) is discussed with the relevant local authority (and in some cases perhaps even the statutory consultees) at an early stage, this helps to speed up the process of environmental assessment and avoid costly delays at a later stage when questions and issues are raised which could well have been dealt with at the outset.

There are procedures for establishing the need for environmental assessment in the case of Schedule 2 projects where there is some doubt, or disagreement between the local planning authority and the developer. For example, it is open to a developer at any time before submitting a planning application to ask the local planning authority for their opinion on whether an environmental statement is required. If an authority gives an opinion that environmental assessment is required, and the developer does not agree, there are procedures enabling the developer to apply to the Secretary of State for a direction. The Secretary of State also has the power to make a direction on his own accord, without any application for a direction from a developer – for example, if he has received representations from third parties. Before making such a direction it is likely that the Secretary of State would give the local planning authority and the developer an opportunity to comment.

The regulations set out "the specified information" which must be provided by the developer in every case. This includes the minimum information required by the Environmental Assessment Directive but goes further than the Directive by requiring the developer to include a description of the likely significant effects, both direct and indirect, of the development on the environment by reference to its effect on the various listed items (which follow the listed items in the Environmental Assessment Directive). The regulations set out a comprehensive list of further information which can be submitted by way of explanation or amplification.

As is apparent from the procedural requirements outlined above, the regulations encourage participation by the public and other relevant interested bodies. In particular, the regulations provide that the Nature Conservancy Council (NCC) and Countryside Commission (now English Nature and the Countryside Commission in England, and the Countryside Council for Wales) must be consulted and given a copy of the environmental statement in all cases where environmental assessment is required. Her Majesty's Inspectorate of Pollution (HMIP) must also be consulted in certain cases. Developers are encouraged to discuss the scope and contents of their environmental statement with the relevant consultees before it is prepared. However, they are not required to do so and a common complaint from, *inter alia*, the NCC (pre-April 1991) and the Countryside Commission about how the environmental assessment procedure is working in practice is the quality of the environmental statements submitted. They consider that this could be improved upon if developers were to consult with them prior to preparation of their environmental statements. Such an approach is worth considering and may well speed up the environmental assessment process, doing away with the need to perhaps provide extra information in answer to a new question raised by a consultee. In some cases, it may also do away with an objection to a project, if a consultee is satisfied that a developer has done his homework and put some effort into the production of an environmental statement.

Whilst generally the statutory consultees will volunteer information and assistance if requested by a developer, there are also mandatory requirements set out in the Planning Regulations whereby the developer can require the statutory consultees to make available any information in their possession which is relevant to the preparation of the environmental statement. The onus is on the developer to inform the various public bodies of the information that he considers helpful to him in the preparation of his environmental statement, and there is no duty on the public bodies to carry out research or to express a view on the merits of the application at that stage. Additionally, there is no obligation to make available confidential information.

It should also be noted that in many cases, particularly those involving complex technical issues, local authorities may not feel competent to appraise an environmental statement and may instruct independent consultants to carry out such work and provide an assessment of the proposals. Whether or not this is the case will obviously depend upon the financial resources of the authority in question. Nevertheless, however, developers should not be surprised if this is the approach adopted by the authority.

It must also be kept in mind that the environmental statement produced may end up at a public inquiry into the planning application for development if, for example, consent is refused by the local authority or the application is not determined in the relevant time frame. Since the role of the environmental statement at the inquiry is to provide the basis for evidence presented by the different witnesses, great care must be taken in its preparation and drafting.

Some practical aspects are worth bearing in mind:

PREPARATION

- Importance of timing, putting the right team together and working early with thoroughness.
- Finding the right person to co-ordinate the project.

PRESENTATION

- Effective presentation by well-qualified and experienced witnesses.
- Any presentation should be: sensible, simple and sensitive.
- Every aspect of environmental impact should at least be considered, even if later dismissed as not being relevant to the particular case.
- Support assertions and conclusions with appropriate evidence.
- Environmental evidence is not different to any other type of evidence, leave no stone unturned.

DoE and other guidance

DoE guidance sets out to explain the Planning Regulations and gives advice upon their implementation, including a detailed commentary on the procedure in Appendix B. Part of the guidance considers how to decide when an environmental assessment is needed for those projects falling within Schedule 2 of the regulations where environmental assessment is only required if "the project is likely to have significant effects on the environment by virtue of factors such as its nature, size or location."

The guidance sets out a list of indicative criteria and thresholds relating to Schedule 2 projects, although it is stressed that these criteria and thresholds are only indicative, and each case will depend upon its facts. Therefore, not every project which is above an indicative threshold will require environmental assessment. Conversely, environmental assessment may be required where a proposal is below an indicative threshold. This guidance should be studied carefully when environmental assessment is proposed.

The DoE has also published an advisory booklet (*Environmental Assessment – A Guide to the Procedures*) setting out more detailed guidance on environmental assessment. The guide is aimed at developers and includes a useful check-list setting out the range of issues which may need to be included in an environmental statement, a list of the statutory consultees, and charts illustrating the main procedural stages. It includes a commentary on those projects which are not subject to planning control, but require environmental assessment under specific regulations, such as highways, afforestation and land drainage improvements.

In addition to the DoE guidance, the Countryside Commission have published a guidance book entitled *Environmental Assessment – The Treatment of Landscape and Countryside Recreation Issues* which is intended to provide advice on the way landscape and countryside recreation issues should be treated within the environmental assessment process. As the Countryside Commission are statutory consultees in all cases, this is an essential guidance booklet for all involved in environmental assessment.

Further guidance would be welcome in a number of areas. First, there is widespread agreement that further guidance (including examples) is needed on how to define "significant environmental effects" for Schedule 2 projects. Additionally, some of the indicative criteria and thresholds in the DoE guidance are ambiguous with the result that there is some inconsistency in application of the Planning Regulations across different local planning authorities.

Secondly, whilst there is considerable guidance on the environmental assessment procedure there is little official guidance on the practicalities of how to prepare an environmental statement. There has been widespread comment that the quality of environmental statements produced to date has been extremely variable, and this might be improved by detailed guidance on the preparation of environmental statements. The DoE has commissioned a consultancy firm – Land Use Consultants – to prepare a guide to good practice in the preparation of environmental statements. At the time of writing the Land Use Consultants' report had not been published but was being considered by and circulated within the DoE.

A third area where further guidance would be welcome relates to environmental assessment and pollution control. The government has recently issued a Consultation Paper concerning the inter-relationship between the pollution authorisation procedures required by HMIP, and the planning process. However, HMIP have not produced any detailed guidance concerning environmental assessment.

Monitoring and research

There have been a number of studies of the Planning Regulations and other environmental assessment regulations and how they are operating in practice. One of the more detailed studies was commissioned by the DoE and carried out by the Environmental Impact Assessment Centre at the University of Manchester in 1991. The research was designed to monitor the operation of environmental assessment procedures under the Planning Regulations from when they were first introduced in July 1988 up to the end of 1989.

The study includes a review of all known environmental statements submitted under the regulations between 15 July 1988 and 31 December 1989 in support of planning applications; a survey of 24 local planning authorities not recorded as having received environmental statements, to see whether they had in fact dealt with planning applications where environmental assessment would have been appropriate; and a detailed analysis of 24 cases where an environmental statement had been submitted. The study set out a number of recommendations for action to improve the implementation of the Environmental Assessment Directive in the United Kingdom.

Recommendations include the need for more guidance, increased consultation and more training of all involved in the environmental assessment process including those engaged in the preparation of environmental statements, and those engaged in evaluating the impact assessment. The report includes a recommendation that a developer should be required to provide three copies (instead of one) of the environmental statement to enable environmental statements to be available for inspection at the DoE in London soon after they have been submitted. This particular recommendation is the only recommendation made in the Environmental Impact Assessment Centre's report which was incorporated in the government's Consultation Paper of 30 June 1992 (*Environmental Assessment and Planning: Extension of Application*). The Consultation Paper indicates that the other recommendations set out in the Environmental Impact Assessment Centre's report are still being considered.

Amendments to the Planning Regulations

The Planning Regulations have been subject to minor amendments but the first substantive amendments to the Planning Regulations were proposed in the Consultation Paper of 30 June 1992. This proposes to extend the projects where environmental assessment is required to trout farming, water treatment plants, wind generators, motorway and other service areas, golf courses (all proposed to be added to Schedule 2), and privately financed toll roads (proposed to be added to Schedule 1). Accordingly it is likely that the categories of projects requiring environmental assessment will be extended in the near future.

The Consultation Paper also proposes an amendment to the procedural requirements of the Planning Regulations requiring any further information requested from a developer over and above that contained in the submitted environmental statement to be submitted to the statutory consultees and the general public in the same way as the original statement.

Controversy over the United Kingdom's implementation of the Environmental Assessment Directive

The United Kingdom's implementation of the Environmental Assessment Directive has given rise to criticism on a number of counts including:

(i) In cases where planning permission is not required it is apparent that a number of projects which are required, under the terms of the Environmental Assessment Directive, to be subject to environmental assessment are slipping through the net. These include, for example, developments covered by crown immunity (although at the time of writing this is under review), development which is permitted by the General Development Order without the need for an application for planning permission, and projects which are specifically excluded from the definition of development in the Town & Country Planning Act 1990 – for example use of land for intensive agriculture.

(ii) Further the Planning Regulations do not actually require the decision-maker to carry out an EIA, but rather the decision-maker is required to consider the environmental statement provided by the developer and the observations made by consultees. It is arguable that this is not a correct implementation of the Environmental Assessment Directive.

(iii) There has been some controversy surrounding the date from which the Directive is to come into force in the United Kingdom. Under the Planning Regulations the environmental assessment provisions only apply to projects where an *application* for planning permission was lodged on or after the date when the Planning Regulations came into force – 15 July 1988. By contrast, the Commission considers that the provisions of the Directive came into force on 3 July 1988, and should apply to all projects for which consent had not yet been granted on 3 July 1988. As there were a number of road schemes in progress in July 1988 this has caused some controversy with the Commission, with the United Kingdom arguing that the Directive did not apply to these projects and the Commission arguing that consent should not be granted until the provisions of the Directive have been fulfilled.

(iv) The relevant decision-makers (usually the local planning authority) are failing from time to time to require an environmental assessment even though projects are likely to give rise to significant effects on the environment, and on other occasions are failing to ensure that all relevant impacts are considered properly in the assessment, or to ensure that the public consultation procedures laid down are observed.

It is possible that the Commission will bring infringement proceedings against the United Kingdom relating to some or all of the above issues.

The relationship of environmental assessment and development plans

As planning law currently stands in the United Kingdom there is no formal environmental assessment procedure required in the preparation of development plans, although such plans must contain proposals for the conservation and improvement of the local environment. It would be logical to extend formal environmental assessment to development plans. This is particularly so in the United Kingdom as the Planning and Compensation Act 1991 inserted a new section 54A into the Town & Country Planning Act 1990 which introduces a presumption in favour of development in accordance with the development plan unless material considerations indicate otherwise. This is indicative of the general move towards plan led development, which is reinforced by the revised DoE Planning Policy Guidance Note 1 (PPG1) – General Policy and Principles, which was published on 10 March 1992. With this change in philosophy it is illogical that formal environmental assessment is required on a site specific basis, but is not required as part of the process for the preparation of the development plan.

The government has not shown any great enthusiasm for an extension of formal environmental assessment to development plans. The Town & Country Planning Association, in its submission on the government's White Paper on the environment, recommended that environmental assessment should be extended to cover development plans; and the energy characteristics of individual schemes and development plan proposals through an energy audit. However the White Paper preferred a more conservative approach stating "EIAs impose costs both on developers who are required to assemble the information and on the authorities which must evaluate it. Any case for the extension of the application of EIA must therefore be considered carefully."

In fact the government has paved the way for extending environmental assessment to any form of development by inserting a clause into the Planning and Compensation Act 1991 enabling the Secretary of State to make regulations requiring an environmental assessment for any form of development. It is likely, however, that this enabling legislation will only be used for minor alterations while any major amendment to the environmental assessment regime in the United Kingdom will be European Community led.

It is almost inevitable that environmental assessment will be extended throughout the European Community in due course to cover plans, policies and programmes, although the precise timescale is difficult to predict. As long ago as 1982 the European Parliament delivered an opinion to the Commission in which it urged the Commission as soon as possible to submit a proposal to extend environmental assessment to public plans and programmes and not just projects. It is understood that at the time of writing the Commission is progressing the development of a proposed Directive on the environmental assessment of policies, plans and programmes.

Trans-boundary issues and Espoo Convention

The Environmental Assessment Directive makes provision for a copy of the environmental statement to be forwarded to any other Member State if the project is likely to have significant effects on the environment in another Member State. This information must be provided either by the Member State where the project is to be located, if they are aware that the project is likely to have significant effects on the environment in another Member State, or as a result of a request from a Member State likely to be affected. The Environmental Assessment Directive envisages that the provision of such information will form the basis for any consultations necessary between the Member States.

On 25 February 1991 the United Nations Convention on Environmental Impact Assessment in a Transboundary Context was signed at Espoo, Finland. The convention has potentially wide implications for transboundary consultation, although it has not yet been adopted by the EC Member States. The requirements of the Espoo Convention are much wider than the Environmental Assessment Directive providing that, in the case of any proposed activity listed in Appendix 1 which is likely to cause a significant adverse transboundary impact, affected parties must be notified. The projects listed in Appendix 1 to the convention are wider than those listed in Annex I to the Environmental Assessment Directive, and include projects such as ground water extraction, petroleum storage facilities and deforestation of large areas. Rather than the somewhat woolly comments about consultation set out in the Environmental Assessment Directive, the convention permits the affected party to elect to participate in the EIA.

The convention lays down minimum information to be included in the EIA documentation (which is more extensive than that required by the Environmental Assessment Directive) and requires a description (where appropriate) of reasonable alternatives (for example, locational or technological) to the proposed activity and also the no-action alternative. The convention gives the public from a country likely to be affected by a project equivalent rights of participation in the EIA procedures to those provided to the nationals of the country where the project is to be located. So, for example, under this provision foreign nationals would have the right to appear at British public inquiries.

The convention provides that EIAs shall as a minimum requirement be undertaken at the project level of the proposed activity. However, it suggests that, to the extent appropriate, the parties shall endeavour to apply the principles of EIA to policies, plans and programmes.

The Convention will enter into force 90 days after 16 countries have ratified it and will require ratification by a sizeable number of EC countries before it takes effect. The European Commission has made a proposal for the European Community and its Member States to ratify the Espoo Convention, but at the time of writing the Espoo Convention had not been adopted by the EC Member States.

The future for environmental assessment

It is not always possible to say with certainty what the future for environmental assessment is likely to be. However in the United Kingdom, in Europe and beyond there is a general trend towards greater openness on environmental issues and a move towards making environmental information publicly available.

The EC Fifth Action Programme on the Environment also places more emphasis on the concept of "shared responsibility", that is, involvement from Central Government, industry and the general public, which again requires greater availability of information (see Chapter 4).

A variety of systems for assessing the environmental effect of projects or industries is also being developed. One new concept which has attracted interest is that of Integral Environmental Planning (IEP) which integrates environmental considerations into planning for industrial and commercial developments. One of the functions of IEP is in the formulation of a risk assessment for industrial activities. The system attempts to quantify the environmental impact that a particular industrial activity may contribute to a particular locality and encourages the formation of what is described as a "buffer zone" between industrial activities and such uses as residential areas which may be adversely affected by such development. The size (and shape) of such a buffer zone would vary, depending upon the type of industrial use in contemplation and its particular characteristics. Whilst IEP is similar in many respects to the system of environmental assessment in place for certain projects in the United Kingdom, IEP is more concerned with the impact of industry on particular areas of land than an assessment of one development project. It may, therefore, apply to whole towns or cities or areas within them. In practical terms, it can, therefore, be described as a system of land-use zoning based upon environmental considerations.

It, therefore, provides another example of how, even before a development is allowed to take place, the acceptability of the proposed development will, in many cases, be measured by reference to its likely impact on the environment. The trend seems to be very much towards an expansion of the range of developments whereby formal assessments of the kinds discussed will be required. In such circumstances, the risk that a formal assessment may become necessary for redevelopment in due course means that at least a cursory, informal assessment should be undertaken before any significant land transaction is entered into: prudence dictates that any matters which may come to affect land values should be identified and inquired into at the earliest opportunity. This notion is nothing new – local authority searches have long been common practice. The message here is simply that the breadth of such enquiries needs to expand. In a slightly different way, contamination issues have come to represent a very real threat not only to the possibility of land development and use but, in the context of the proposed registers, very directly to maintaining the value of land as an investment.

Contaminated land and its implications

It is possible for an innocent purchaser of land to be held liable for activities which took place prior to that acquisition. Thus someone buying contaminated land may be held liable for the cost of cleaning up that land, even though he took no part in and had no knowledge of the activities which caused the contamination in question.

The most important thing to bear in mind when acquiring land is the principle *caveat emptor* or let the buyer beware. The purchaser of land must satisfy himself as to the possible presence of contamination, which may give rise to financial or other liabilities. A potential purchaser should, therefore, make enquiries about possible environmental liabilities at the earliest opportunity in negotiations, particularly if the land has been used previously for an industrial purpose. These enquiries may take many forms, ranging from careful examination of historical records of consent registers (maintained by planning and pollution control agencies), scanning of newspapers and trade journals for indications of past pollution incidents, and site investigations and enquiries of local people.

Because of this risk of undisclosed liabilities on the purchaser, pre-contract enquiries are used to request information about the property. Standard pre-contract enquiries are, however, rarely sufficient where contaminated land or other environmental problems are concerned. The types of questions asked will vary depending on the type of land being acquired and the use to which it is presently put. However, requests should be made for copies of all authorisations and other documents such as waste disposal contracts, health and safety policies and environmental audits conducted, together with advice of any enforcement action being taken.

It may also be desirable to conduct on-site investigations, often referred to as acquisition audits. This may involve sampling of soil, water and the like on and around the site, surveys and inspections by suitably qualified inspectors, and interviews with people who have worked at the site in order to learn about its history.

Having obtained information about the site, steps can be taken to reduce the risks which such information has shown to exist and to allocate responsibilities consciously between vendor and purchaser. This may include securing appropriate contractual protection from relevant parties, such as warranties from vendors that the land has not been used to store certain substances or used for certain contaminative purposes. (The significance of warranties and indemnities is considered in Chapter 9.)

It may perhaps be appropriate to seek "green warranties" from a vendor. Such warranties are a contractual obligation on a vendor. They may cover matters such as the condition of land being purchased and, depending on the past or present use of the site, can cover a wide range of matters.

It may be useful to obtain a warranty that any site activities comply with the relevant legislation or codes of practice, and that there are no actions pending for breach of such legislation. Warranties should also cover insurance, assuring a purchaser that insurance is in force and has been properly maintained on the property.

It also might be appropriate to obtain insurance cover if it is available. Insurance cover should be considered not only for risks which may be innocently acquired on purchase of a site but also for liabilities which may be incurred as a result of what happens after a site has been acquired.

Registers of contaminative uses of land

Section 143 of the Environmental Protection Act 1990 requires a Council to produce and maintain a public register of land which may be contaminated by specific uses. One of the main functions of such a register would be to notify purchasers of land that possible contamination of a site had occurred. It was originally proposed that regulations relating to this duty would come into force in early 1992. However, as a result of a consultation exercise undertaken by the DoE, which resulted in significant objection to the possibility of planning "blight", the DoE did not announce the publication of draft regulations until July 1992.

The problems envisaged with the compiling of such registers were that land placed on the register and land in its vicinity would become subject to planning "blight" upon its use, and that the value of it and other land in the area would be affected. The fact that land would appear on a register would not necessarily imply that it actually was contaminated; it would merely show that contaminative uses had been carried out with the possibility of actual contamination. If land did not appear on the register, this would not necessarily guarantee that it had not been contaminated. Moreover, the presence of natural contaminants such as radon gas would not be noted in the register. It was also suggested that further investigations of such land would need to be undertaken to determine whether or not it was subject to actual contamination. The government, industry and landowners were, however, unable to decide who should bear responsibility for such investigations. The effect on asset values is discussed further elsewhere in this chapter.

After a delay of some months, the DoE issued revised proposals for the registers. The proposals are not, however, that dissimilar to those which preceded them. The number of contaminative uses to be included in the registers has definitely been reduced. The lengthy list released by the DoE in 1991 has been cut to eight items which include sites used for the manufacture of gas, coke, lead, steel, asbestos products, chemical products and waste disposal sites.

A register does not have to be compiled in any particular way, but the chosen format must allow members of the public to identify readily information about any particular site. The register will now be divided into two parts: A and B. Part A will comprise land which is subject or has been subject to a contaminative use and which has not been investigated or treated in any way. Part B will comprise land which has been subject to investigation and/or some kind of treatment. A description of the work carried out will be provided.

The register will, therefore, put prospective purchasers on notice that land has been subject to a contaminative use of the kind described in the regulations. Part B of the register is intended to help purchasers of land obtain more information about it so that they can assess whether or not the treatment described in the register has been effective. However, as before, entry on the register will not be proof of actual contamination. The registers do not attempt to give an assessment of the present condition of any land.

One particular problem with the regulations is that they do not give a right to appeal against entry of land onto the register; rather, they provide that two months notice of proposed entry must be given to the owner or occupier of the land and any person with an interest in the land must be given an opportunity to inspect the particulars to be inserted on the register. It would, therefore, be necessary to bring an application for judicial review to challenge the local planning authority's decision to place land on the register. It is unlikely that such an application would be a very effective method of challenging a decision. Alternatively, a complaint may be made to the Local Authority Ombudsman. However, the Ombudsman only has power of recommendation and the Local Authority is not obliged to accept his decision. It would be preferable to have some kind of appeal procedure inserted into the regulations. Even if land has been treated and all contamination removed, there is still no provision in the regulations to remove that land from the register. No doubt there will be further debate in connection with this very sensitive topic.

Development of contaminated land

The DoE has made it clear that contaminated sites may require remedial action to be taken before any subsequent development can take place. It is, therefore, likely to be more expensive to develop contaminated land and this expense must be taken into account when any project is costed.

DoE Circular 21/87 and its draft Planning Policy Guidance Note on Planning and Pollution Controls (intended to replace Circular 21/87), deals with the development of contaminated land. The circular makes it clear that, in the interest of making full and effective use of land, a balance must be struck between the hazard of using contaminated land and the need to bring such land back into beneficial use. The responsibility for assessment rests with a developer.

The DoE's approach is that, whilst few sites are so badly contaminated that they cannot be re-used at all, the types of use permitted on land may well be restricted by contamination. Each application would, therefore, have to be considered on its own merits. If remedial action is required, guidance makes it clear that planning permission should be granted subject to conditions specifying remedial measures to be carried out.

The impact of ground contamination legislation on asset values in the UK property market

Earlier sections of this chapter have identified the way in which increasing obligations and liabilities are being placed on owners and users of land through legislation which is intended to deal with pollution in its many and varied forms. To some extent these obligations are not new, but standards are becoming more stringent and, significantly, public awareness is increasing.

Property values derive from an interaction in the market place between supply and demand. The market operates with considerable freedom, but it does so within the framework of public attitudes and legislative controls. Over the post-war period public involvement in the private use of land has expanded widely.

Restrictions on the use and design of land and buildings are imposed through planning controls, and concern for the safety and health of occupants is recognised in health and safety legislation. All of these controls have potential cost implications. For that reason they impact on the nature and cost of buildings in use and the standards required in development. Costs are an essential element in determining asset values.

The impact of controls on property values in some areas is tried and tested. For example, the market has long experience of the wide scope of planning controls, and the fact that these controls determine the way in which property values are distributed between urban and rural areas and within regions and districts. In other respects, however, public and, indeed professional, awareness is in its infancy and growing in response to the widening impact of legislation.

Ground contamination is an area where at the present time knowledge and experience is only gradually emerging, though US practice appears to be ahead of that in the United Kingdom. It is, however, not the first time that the property market has had to come to terms with new liabilities which have potential for influencing value. Within the context of legislation, perhaps the most important statute which affected asset values was the system of planning control established by the Town & Country Planning Act 1947.

In many respects the pressure for control which existed at that time was politically led. Concern for the environment and the risks associated with ground contamination is, however, the result of public pressure, and in measuring its impact on property values a closer parallel from a valuer's viewpoint was the widespread concern which arose some 20 years ago in relation to the use of high alumina cement concrete (HAC) as a component in building construction. Concern was not prompted by statute, but arose in the wake of extensive publicity which followed the collapse of a relatively small number of buildings where a combination of physical circumstances and use of the material produced serious structural failure. Almost overnight, buildings which contained HAC became unsaleable, valuations were qualified and were prepared on the assumption that the material had not been used in construction. More often than not buildings were blighted without any distinction being drawn between those buildings where the components using HAC had been positioned in such a way as to pose a potential risk of structural failure, and those where the risk was remote. Over time, however, informed purchasers came forward who were prepared to investigate the problem on an individual basis. Many cases arose of favourable purchases with considerable potential for profit.

The response of the property market to the liabilities associated with ground contamination, in particular the proposed "registers" of potentially contaminated land, is in many ways similar to initial reaction to the problem of HAC. The major difference as far as ground contamination is concerned, however, is that potentially very large areas of land, including significant proportions of urban areas, could be affected. Frequently contamination will not be immediately apparent without careful investigation and the cost of a failure to identify a heavily polluted site could be considerable. A property acquired as an asset could turn out to be a liability. At least with HAC the demolition of the building would eradicate the problem. With ground contamination on the other hand, demolition of the structure may be nothing more than the first in a whole series of complicated and costly procedures which may be required before the problem has been eradicated and development can take place. Removal of the top soil may not be sufficient if ground water is being or has been affected.

The effect of blight, regardless of the source from which it springs, is invariably more far reaching in its impact than is justified on a rational analysis of the risk which is involved. Over recent months considerable publicity has been given to decisions by prospective purchasers to withdraw from acquisitions ostensibly on the ground of concern in relation to contamination. In all probability it may have been but one of a number of factors. It seems to be the case, however, that the traditionally cautious attitude of the institutions has resulted in withdrawals from negotiation in some cases once a possible risk of ground contamination has been identified. Initially withdrawal was typically total. However, even at this early stage it is apparent that the same institutions are beginning to look beyond the problem of general blight to the potential risks, costs and liabilities which a particular site could involve. The extent to which a pragmatic view can be taken will depend to some extent on the attitude of banks and investors in funding such acquisitions.

The way in which these investigations can be carried out has already been referred to. In so far as it is possible to draw any conclusions at this stage, it is evident that where contamination is physically confined and can be adequately treated the cost involved (together with an allowance for possible uncertainty) provides a framework within which re-negotiation of the purchase price can take place, without the necessity for a sale being aborted. On the other hand if the contamination is affecting ground water, with the attendant risk of migration off site, the consequent liabilities, the difficulty of assessing the scale of the problem or the ultimate extent of the liability, present at the moment a risk which many purchasers in the market are unwilling to take.

A question which is frequently asked is whether the proposed registers of potentially contaminated land are creating the problem, or whether ultimately they may produce a means of confining blight. To an extent registers have done their job, by increasing general awareness of the issue. They are, however, rapidly becoming irrelevant, as solicitors now make detailed enquiries on property transfers, and banks and insurers require assurance on lending decisions as environmental costs begin to have greater effects on profits and share value.

Blight is not a consequence of the proposed registers. It is the result of increasing public and corporate awareness of the costs which contamination can impose. Statutory controls represent only one strand of the problem. Developers have to satisfy not only Central and Local Government in relation to their proposals, but ultimately they have to provide land and buildings which occupiers wish to use and are prepared to acquire and pay for. A number of companies, particularly those based in the United States, are unwilling to be associated with the use of contaminated land. To such occupiers the Registers are no more than a means of initial investigation into whether a particular site is likely to prove acceptable.

The concern of government over the risk of widespread blight is understandable and there can be sympathy with the objective which was in proposals to restrict the scope of the registers. However, the government cannot reverse the trend towards public awareness and recognition of public concern over contamination. The proposal to restrict the scope of the registers to sites "with a high probability" of contamination, simply begs the question of lower probability sites. Many sites with possibly significant contamination costs (*e.g.* shipyards, tanneries) have been omitted. The new draft states that there will be a review of the registers in the light of experience. In other words if a site may be contaminated, even though it is not on the registers today, it could be tomorrow. Blight is only partly about known liabilities. A far greater concern arises through uncertainty, and the way in which the registers are now to operate has, if anything, exacerbated the problem.

So far the impact on property values of liabilities in relation to ground contamination has been muted. Valuers are being advised by the Royal Institution of Chartered Surveyors to issue valuations on the assumption that no investigation has been made in relation to ground contamination, or that the extent to which any contamination which might be found could lead to the property being entered on the contamination land registers. Banks on the other hand are more cautious and many are insisting that valuations for loan purposes will now only be accepted if accompanied by an environmental audit and investigation.

At the present time the property market is severely depressed, and the unwillingness of banks to lend on potentially contaminated sites is of little consequence when lending policies are, for quite separate reasons, dictating that funds for property development are not in any event available. As the market recovers, however, it can be expected that the issue of ground contamination is going to have an increasingly widespread effect on property values. It is only a matter of time before the issue of qualified valuations will become unacceptable and it seems probable that an investigation by some means or another into ground conditions will become as essential as a structural survey in all property transactions.

Effect of contamination on land value

In terms of effects on site value, there are two elements:

(i) The actual cost of cleaning up the site to a condition appropriate for its intended use, and avoiding effects on neighbouring land uses or the wider environment (in particular, surface and groundwaters).

(ii) The instantaneous effect on land value caused by the market perception of a site's appearance, or potential appearance, on a register (or simply the attachment of a "potentially contaminated" label).

The former can be estimated with reasonable confidence, given appropriate site investigation results and a knowledge of the geology and hydrogeology. For example, it is possible to cost soil excavation and removal, disposal (landfill or treatment), on-site biological treatment, vacuum extraction for groundwater clean-up etc. These costs are often considerable.

The clean-up costs often represent the bottom-line in price negotiations for property acquisition. The effect of the "potentially contaminated" label, *per se*, is much more difficult to assess. In some ways, if the original Register proposal, with its much wider scope of contaminative uses (including, for example, docks, railway land, shipyards, mining etc) had gone ahead, it would have had the effect of maintaining land values by virtue of the fact that, in many areas, the land registered would comprise very large proportions of that available. For example, it had been estimated that 35% of Birmingham would be on the register. Very little evidence of the effects of a contaminative history on property value is publicly available. Currently, in price negotiations, reductions of 10-30% are commonplace.

Over the medium-term, however, buyers and investors alike will seek greater certainty over the issue, and will be less willing to "take a view" on the basis of a paucity of site-specific information.

Uncertainties

A major problem with attempting to estimate the cost of clean-up, is the high degree of uncertainty involved, in terms of, for example:

- the definition of acceptable clean-up standards; and
- the choice of a clean-up methodology.

There are no fixed standards in the United Kingdom below which a site can be classified as "clean". The determination of clean-up targets is site specific, and in most cases involves negotiation with the regulatory authorities. Clearly, such negotiations cannot fit easily into the commercial timetables associated with property transfers, and may be impossible if a transaction is confidential. However, the clean-up target is a critical factor in estimating costs.

Future developments

The government has also announced its intention to set up an urban regeneration agency to deal with some 150,000 acres of vacant land in the country's towns and cities. In its Consultation Paper on the subject issued at the end of July 1992, and its Bill issued in October, the government proposes giving responsibility for city grant and derelict land grant schemes in urban and non-urban areas to the new agency, together with responsibility for English Estates, a non-departmental public body which is involved in factory building in urban areas.

One of the agency's key tasks would be to enable vacant and derelict land in urban areas to be brought back into beneficial use. The intention is to assist the private sector in developing land as well as working with developers and local authorities in identifying sites and reclaiming them. The agency would be given power to compulsorily acquire land where owners refused to regenerate it or to sell the land. However, responsibility for Urban Development Corporations (UDC) will be retained by the DoE and the urban regeneration agency would not normally have a role in UDC areas.

It is commonly thought that between 50,000 and 100,000 sites covering in the region of 50,000 hectares may be subject to contamination and that many of these sites will require significant remedial work before they can be used beneficially. Site assessment costs alone run to something like £15,000 per hectare and estimates for clean-up run into billions of pounds.

Even if these estimates prove to be exaggerated, contaminated land will still be one environmental issue which is here to stay. The United Kingdom has, up until now, been fortunate in that it has not suffered a contaminated land problem to quite the same extent as the problem experienced in the United States with incidents such as Love Canal. It may be that, if a similar disaster were to occur in the United Kingdom, some sort of "superfund" scheme would be required. Figures such as those mentioned above show some of the problems faced by government in its role of encouraging landowners and the industrial sector to come to grips with the issue of contaminated land. The government, faced with a call for increased public accountability for contamination, is trying to encourage industry to be accountable and to bear the cost of cleaning up land, but it also has a responsibility to keep businesses afloat. Whatever happens, businesses should not become so burdened with expenditure that their asset base is constantly erroded. This is a fine line to tread but, as discussed earlier, it is impossible to reverse the trend towards public awareness and recognition of concern over contamination. It is for these reasons, in particular, that one can envisage the growing significance of Eco-audits in relation to land. The scheme, albeit voluntary, would assist owners in understanding the problems of contamination and hopefully move towards a more considered view of protecting the environment by a technical and more commercially sound clean-up programme; and, of course, a greater emphasis on preventive measures.

CONTENTS OF CHAPTER 8

Environmental Liability Issues: Companies, Directors and Auditors

Chapter 8
Environmental Liability Issues: Companies, Directors and Auditors

Although the growing significance of environmental management systems is to be expected, inevitably problems and unforeseen risks can arise. Having said that, if management systems are properly in place, to the satisfaction of the directors, employees, suppliers, customers and shareholders, one would hope that no major difficulty ought to occur. However, in this chapter some of the problem areas that can cause difficulty and exposure to liabilities when adopting and administering environmental management systems are considered from the point of view of the shareholder, the director, the environmental director, the environmental auditor, be it to the client or to third parties.

At the end of Chapter 3 the potential problem was raised of a claim by shareholders who had purchased shares based on what transpires to be a half-hearted implementation of an environmental policy. Would such a claim succeed in the United Kingdom? What other claims involving directors, environmental directors and environmental auditors are likely to arise? This chapter seeks to answer these two questions.

Would a claim by shareholders succeed?

Shareholders' claims

There is a trend in legislation and court decisions dealing with a company's financial management to require directors to look beyond the company's duty to its shareholders and consider the impact of its dealings upon employees, creditors and others. This trend also appears in health and safety legislation. Its application to the environmental activities of companies can be anticipated. It will be essential for directors to be able to demonstrate that they have behaved "responsibly".

There is a close analogy between the developing environmental auditing regime and the existing regime for the financial audit of companies. The overriding requirement to give a "true and fair view" of the state of a company's finances may find its equivalent in a requirement to give a "true and fair view" of a company's environmental impact.

In the current state of the law, the likelihood must be that it would not, unless the practice falls foul of the prospectus rules which will be referred to below. In most circumstances, directors of a company do not have a duty to persons who are, or who might become, shareholders in that company, but only to the company itself. Of course, the company may, in general meeting, dismiss a board which it thinks has been ineffective in carrying out the company's policy but individual shareholders cannot take action in their own right.

One possible exception is where a statement of environmental policy is contained in a prospectus or listing particulars. These are documents which are required to be made available to the public in connection with an issue of shares in a public company and which contain various statements about the company's affairs. The information which they have to contain is defined at some length in the appropriate legislation and the rules of the London Stock Exchange, but in addition to the information specified in those sources, the listing particulars are required to "contain all such information as investors and their investments advisers would reasonably require, and reasonably expect to find there, for the purpose of making an informed assessment of . . . the assets and liabilities, financial position, profits and losses and prospects of the issuer of the security . . ." (Section 146(1)Financial Services Act 1986). Breach of this clause can give rise to liability to investors on the part of any person "responsible" for the particulars. This latter category includes the company itself, its directors at the time of the preparation of the particulars and any future directors named in the particulars and in certain cases, experts and others whose opinions are quoted in the prospectus with their consent.

It seems reasonable to assume that the fact that a company has subscribed to the Eco-audit scheme and has thus submitted the environmental management of certain of its sites to public examination is a fact which should be disclosed in the listing particulars. To the extent that the financial commitment involved in satisfactorily running the environmental management system could affect the financial prospects of the company, it is arguable that an assessment of its financial impact would also form a disclosable item.

Whether a proposed change in the company's policy could involve any liability to a shareholder who had taken shares in reliance on the company's environmental commitment is a more difficult question. If the prospectus stated that the board was committed to membership of the Eco-audit scheme at a time when the directors had already taken a decision not to prepare the next environmental statement, then there would be a strong argument that the prospectus was misleading and that the shareholder who had based his decision to buy shares upon the company's environmental policy would be entitled to relief – in most cases, to have his shares bought from him. If, as is more likely, the statements in the prospectus are true at the time they are made but a change in factors occurring shortly afterwards – for example, an environmental audit of a particular site which shows that the cost of dealing with contamination is 10 times what the directors anticipated – resulting in a subsequent change of policy on the part of the board would normally not give rise to any right of action. The greater the period of time between the issue of the prospectus and any change of policy, the less likelihood there is of any claim being successful.

Directors' liability

Directors contemplating introducing an environmental management policy could well ask themselves whether an incidental effect of doing so will be to increase or decrease their risk of their being held personally liable for acts or omissions committed in the course of their employment.

Liability on the part of a director can arise to:

(i) the company of which he is a director;
(ii) the shareholders of that company;
(iii) the creditors of that company; and
(iv) the State or other persons entitled to claim under statute or for breach of statutory duty.

Duty to the company

The basic rule is that a director owes a duty to the company of which he is a director to act in its interest to the exclusion of the interests of third parties. As discussed in Chapter 3, this can cause difficulties for the altruistic director where he commits the company to some course of action which is not in its interest. However, the test which directors have to face is not a particularly onerous one. They have to genuinely believe that the acts which they carry out are for the good of the company and that belief must be one which an honest director could reasonably hold. A director who genuinely believes that the adoption of an environmental management policy will benefit the company in the short- or medium-term is unlikely to be acting in breach of his duty. As mentioned in Chapter 3, the position would be different if the adoption of an environmental management policy was not permitted by the memorandum of association of the company. It is difficult to imagine circumstances in which this will be the case, particularly where a large part of the policy is concerned with compliance with laws that are either in existence or likely to come into existence in the near future. It is only if the proposed action by the directors goes beyond this, so that they could be said to be motivated more by concern for the environment than concern for the well-being of the company, that difficulties might arise. It is unlikely that the adoption of an environmental management system could provide evidence of such altruism although individual acts carried out under the ambit of an environmental management system could do so. On the whole, therefore, it is unlikely that directors would incur liability to the company for introducing an environmental management system. It is, indeed, considerably more likely that they could risk committing a breach of duty by failing to introduce such a system.

It comes as a surprise to many people to know that there is no general statement of a director's duties anywhere in the Companies Acts or any other UK statute. Some particular duties are statutory but most are derived from judgments of the courts over the last 150 years. An oft repeated definition of a director's duties is contained in the judgment of Mr Justice Neville in *Re Brazilian Rubber Plantations and Estates Limited (1911)* where he said that an executive director must display:

> "Reasonable care in [the despatch] of the business of the company. Such reasonable care must be measured by the care an ordinary man might be expected to take in the same circumstances on his own behalf. He is clearly ... not responsible for damage occasioned by errors of judgment."

Breach of duty or misfeasance arises when a director fails to display the care which an ordinary man might be expected to take on his own behalf. A good example is presented by the case of a director who signed blank cheques and handed them to an unsupervised employee. Needless to say, the employee defrauded the company of a large amount of money. The director claimed that he was completely innocent of the fraud (which was true) and that he had no reason not to trust the employee (which was probably true). The court replied, however, that if it had been the director's own money instead of the company's he would not have handed over a blank cheque without further investigation and, therefore, concluded that he had acted in breach of duty.

Signature on a blank cheque is an extreme example of lack of management control, but most breach of duty cases which do not involve dishonesty or conflict of interest (neither of which are of particular relevance to our present subject) centre on a lack of management control. Indeed, when we come to look at the question of statutory duties imposed on directors, the need for management control and regular review of the company's affairs is the motivation behind much of the legislation. This can be seen, for example, in the case of directors accused of wrongful trading under the current insolvency legislation. It used to be an adequate defence under previous insolvency legislation to state that the director in question did not know the company was in financial difficulties because the last audited accounts had been prepared nine months before and the company had been viable then. That defence is no longer available. The director will now be liable if he knew or "ought to have concluded" that there was no reasonable prospect of the company avoiding insolvency. The important words are "ought to have concluded" because section 214 Insolvency Act 1986 goes on to say that:

> ". . . the facts which a director of a company ought to know or ascertain, the conclusions which he ought to reach and the steps which he ought to take are those which would be known or ascertained, or reached or taken, by a reasonably diligent person having both:
> (a) The general knowledge, skill and experience that may reasonably be expected of a person carrying out the same functions as are carried out by that director in relation to the company, and
> (b) The general knowledge, skill and experience that that director has."

To avoid the ingenious defence that the negligent director did not carry out any functions in relation to the company at all, the section provides that references to functions carried out by a director include "functions which he does not carry out but which have been entrusted to him."

The provisions have been set out at some length because, although they relate only to the offence of wrongful trading, they probably reflect the developing modern attitude of Parliament, and to a lesser extent, the courts, to directors' duties. The director who claims that the company was solvent when he last looked at its audited accounts is now likely to be met with the reply that a reasonable person in his position would have ensured that the company prepared regular management accounts from which its financial position was ascertainable on a monthly basis. Although it is rare for the question of the efficiency of directors to be raised in court, except in the case of insolvency, it cannot be assumed that the "ordinary man in the conduct of his own affairs" test is the only test which is now relevant.

The significance of these remarks to the adoption of an environmental management policy should be clear. The director who says that he did not know that the company was breaching emission limits because the company only tested emission limits once a year and they were satisfactory when last tested, is in very much the same position as the director who only ever looks at the *annual* accounts. A reasonable director, knowing that the directors are under certain obligations if the company is unable to pay its debts, will ensure that there is a financial reporting system in place to enable him to check the company's solvency at regular intervals. A director who knows the company's statutory obligation in respect of environmental damage will ensure that there is a system in place to monitor environmental impact on a regular basis.

It is probably the case that whilst environmental audits and environmental management policies are comparatively rare, failure to institute an audit or policy, except where specifically required by law, cannot be said to be a breach of duty. There may be exceptions even at this stage, however, if the company's business involves a particularly high degree of environmental risk. This could also be the case if the company owned an asset which was especially hazardous. For example, if a piece of land was known to be contaminated with a substance which was liable to seep into neighbouring land or water courses it is arguable that the directors of the company would be in breach of their duty if they failed to take steps to monitor seepage from time to time.

As environmental management becomes more common, however, the general duty on directors will increase. To use the wording of the wrongful trading provisions, "the general knowledge, skill and experience that may reasonably be expected" of directors in relation to environmental matters will increase. This means that even a voluntary scheme such as the Eco-audit scheme can have an effect on the liability of directors whose companies are not members of that scheme, by increasing the general level of environmental "responsibility" expected from directors.

It is worth emphasising that all of the above comments relate to the duties of a director to the company of which he is an officer. As mentioned above, cases of breach of duty against directors rarely come to court unless they involve dishonesty or conflict of interest. It is suggested that there are three main reasons for this. The first is that directors who have not been involved in dishonesty or a conflict of interest rarely have the resources to meet a claim brought against them by the company. The second is that the claim would normally be instituted by the company through its directors and it is probably not too cynical to suggest that directors are slightly averse to suing their predecessors for what is effectively negligence (albeit a rather extreme form of negligence). There is always a risk that the action might be taken as a precedent by their successors in their turn. Thirdly, the reason that the directors have taken the action complained of – or more likely, failed to take a particular action which they should have taken – is often a wish not to spend the company's money.

Nevertheless, there are circumstances in which directors have been sued for breach of duty and the possibility should not be overlooked. Imagine that a board has ignored various environmental danger signs in order to avoid a financial commitment that would severely affect the company's balance sheet. An environmental catastrophe duly occurs which results in a massive public outcry against the company. The board who were responsible for ignoring the warning signs is sacked. Is it inconceivable that the company, in order to deflect criticism from itself and to indicate that the catastrophe was an aberration caused by the negligence of certain individuals, might sue the directors for breach of duty? The moral is that the prudent director will insist on the adoption of an environmental management policy.

The environmental director

A company which institutes an environmental management policy is likely to give special responsibility for that policy to a member of the board and to have an "environmental director" in the same way that it currently has a finance director or sales director. Is an environmental director at more risk of personal liability than an ordinary director in relation to his duties to the company?

Once again, the primary statement of the law is found in an old case – *Re City Equitable Fire Insurance Company (1925)* – where Mr Justice Romer said that "In respect of all duties that, having regard to the exigencies of business ... may properly be left to some other official, a director is, in the absence of grounds for suspicion, justified in trusting that official to perform such duties honestly." This is not a surprising position. There would be little purpose in appointing one of the directors as a finance director if the other directors were under a duty to monitor the finances of the company with the same diligence as the finance director, although as we have seen, the appointment of a financial director will not discharge the other directors from taking a reasonable interest in the finances of the company.

The same can be said of an environmental director. If the board as a whole sets up an environmental management policy and appoints an environmental director to take charge of its administration, the major duty on other members of the board will be to ensure that regular and satisfactory reports are received from the environmental director and that the environmental policy shows every sign of working correctly. So long as there is a system in place by which emissions can be measured, for example, the other directors will not be responsible for measuring the emissions themselves. To that extent, therefore, the appointment of an environmental director and the setting up of an environmental policy could be said to relieve the other directors from a part of their duty or, at least, to define their duty more restrictively.

What of the environmental director himself? There is no question that, even in the absence of any specific provisions in his service contract, he will be expected to have a higher degree of responsibility for matters relating to the environment than the other directors on the board. In this respect the wrongful trading provisions are noteworthy, with their reference to "The general knowledge, skill and experience that may reasonably be expected of a person carrying out the same functions as are carried out by that director in relation to the company." To the extent that a director can have liability that goes beyond the "ordinary man" test, the environmental director may reasonably be expected to have greater knowledge, skill and experience in environmental matters than an ordinary director. The duty may well be greater if the director in question has a background in environmental affairs. Again, reference can be made to the wrongful trading provisions where one of the tests of what the director ought to have known is "The general knowledge, skill and experience that that director has." Thus it is no excuse for an environmental director to say that generally the standard of knowledge among environmental directors is very poor if his own standard of knowledge is considerably above the average. Conversely a director forced into the environmental role against his will cannot plead in his defence a total ignorance of environmental matters if the general standard of knowledge among persons in a similar position is comparatively high. Although the test applied in wrongful trading provisions is not strictly applicable in any other field, there is reason to believe that it represents the direction in which the law is moving.

Shareholders and creditors

The above paragraphs have dealt with the duties which directors might owe to the company itself in connection with an environmental management policy. The circumstances in which directors can owe a duty to a company's shareholders were dealt with at the beginning of this chapter and the circumstances in which such as duty can be owed to creditors are restricted. As such liability is unusual and is no more likely to arise in connection with environmental matters than in connection with any other type of matter, it is not considered further in this section. In almost all cases where directors face potential liability to third parties the source of that liability will be found in statute.

Directors' and officers' personal liability

A development which will be of particular concern to directors and managers is the increasing tendency of environmental legislation to contain provisions imposing personal criminal liability upon the directors and managers etc of companies which have been convicted of an environmental offence. For example, in the United Kingdom under section 157 of the Environmental Protection Act, which is already in force, where a company has committed an offence under the Act "and it is proved to have been committed with the consent or connivance of, or to have been attributable to any neglect on the part of, any Director, Manager, Secretary or other similar officer of the body corporate or a person who is purporting to act in any such capacity" that person may also be guilty of the offence and be liable to be punished accordingly. Similar provisions appear in the Water Act 1989 in connection with offences relating to the pollution of water, in the Control of Pollution Act 1974 which deals with the disposal of waste on land and, most notably, in the Health and Safety at Work etc Act 1974. It is important to note the precondition for liability namely, consent, connivance or neglect.

The meaning of consent and connivance in this context were described in the case of *Huckerby* v *Elliott* as follows:

> "It would seem that where a Director consents to the commission of an offence by his company, he is well aware of what is going on and agrees to it.
> Where he [the Director] connives at the offence commited by the company he is equally well aware of what is going on but his agreement is tacit, not actively encouraging what happens but letting it continue and saying nothing about it."

Neglect is the widest of the three terms. No absolute duty is imposed on the director and some act or omission amounting to neglect must be shown.

Whilst no director has been convicted of an "environmental" offence to date, there have been convictions under health and safety legislation and, given the current profile of environmental issues, it may not be too long before environmental law sees a similar development.

Section 157 of the Environmental Protection Act in addition to providing for personal liability where offences are committed by a body corporate, also provides that where the affairs of a body corporate are managed by its members, a member exercising management functions may be personally liable for offences committed by that body corporate in the same circumstances in which a director etc could have been liable. This provision would seem to pierce further the already much punctured corporate veil and it remains to be seen to what extent the courts will interpret this section so as to impose liability on a parent company for the offences commited by its subsidiary.

There are also implications for corporate acquisitions. The law of conveyancing in England is such that a vendor has only a very limited duty of disclosure and so long as he does not commit any fraud or make any misrepresentation, a purchaser will have no legal redress against him. Furthermore, a vendor is under no duty to respond to a purchaser's enquiries, providing that failure to respond does not amount to misrepresentation.

It is fundamental, therefore, that in virtually every commercial property transaction or business acquisition of consequence, the purchaser should consider and if necessary specifically address environmental issues. First, attention should be given to identification of potential environmental liabilities. Here, the question of whether or not to conduct an environmental audit should be addressed.

Secondly, having identified potential liabilities associated with the proposed acquisition, a purchaser must decide how to structure the transaction to reduce its potential risk under environmental law. There are a large number of possibilities to consider here which apply equally to circumstances where the problem is some other potential liability rather than an environmental one. Accordingly, they will be familiar to those who are regularly engaged in acquisitions and disposals. Suffice it to say that consideration should be given to the following, where appropriate:

(i) Proceeding by way of asset purchase rather than share purchase so that assets to which a potentially high environmental liability is attached can be excluded from the purchase.
(ii) Contractual protection in the form of warranties and indemnities (see Chapter 9).
(iii) Using a subsidiary company to make the acquisition so that the rest of the assets will be protected should the potential liability be realised (note however, that increasingly, the corporate veil may be lifted, for example under section 157 Environmental Protection Act as mentioned earlier).
(iv) Using supply contracts or contracts for services to achieve the commercial objective rather than making an acquisition.

Liability of environmental auditors

Environmental auditors may find themselves liable to claims by third parties on three different bases:

(i) in contract, if they fail to carry out the terms of their appointment with the company;
(ii) in negligence to the company or a third party; and
(iii) under any statute, Directive or other regulations affecting their liability.

Liability to the company in contract and negligence

The potential liability of financial auditors, and other professional advisers, in contract and negligence has been the subject of considerable examination in the courts over the past few years. As we will see, the analogy between the role of a financial auditor and the role of an environmental auditor is comparatively close and, notwithstanding the absence of any regulations or cases about the duties of environmental auditors, one can use that analogy to discuss the position of an environmental auditor.

Auditors brought in to audit a company's accounts are engaged by the company under a contract, normally on the standard terms of the firm of accountants in question. The contract normally contains various exclusion clauses and, in particular, attempts to set out the extent to which the auditors will be entitled to rely on the accuracy of information supplied by the directors. "Attempts" is used here because the efficacy of such clauses may be called into question when they conflict with the statutory duty imposed upon auditors by the Companies Acts (see below).

Auditors also owe a duty of care to the company which they audit. That duty requires them to carry out the audit using professional skill and care. For practical purposes the duty is virtually indistinguishable from the normal contractual duty and will not be discussed further in this section – its importance emerges when considering auditors' liability to parties with whom they are not in contract.

If auditors fail to carry out an audit in accordance with their contract – if, for example, they fail to notice and report upon discrepancies in the accounts which an auditor should, on a reasonable view, have noticed – they incur potential liability to the company for breach of contract. The question which forms the subject of debate is what, on a reasonable view, an auditor should notice – in particular, how thoroughly is he obliged to investigate the company's affairs? To discuss the matter further it is necessary to look at what the Companies Act (CA) says about the function of financial auditors.

It is the duty of the *directors* of a company to keep proper accounting records (s 221 CA 1985) and to prepare annual accounts which comply with the Companies Act. Those accounts must give a true and fair view of the profit and loss of the company for the period in question and of the state of its affairs at the end of that period (ss 226 and 227 CA 1985).

The duty of the *auditors* is to report to the members of the company on the annual accounts stating whether they have been properly prepared by the directors in accordance with the Companies Act and whether they show a true and fair view. In doing so they have to "carry out such investigations as will enable them to form an opinion as to . . . whether proper accounting records have been kept ... and whether the company's individual accounts are in agreement with [its] accounting records" (ss 235 and 237 CA 1985). Put simply, it is the directors' responsibility to keep the records and prepare the accounts and the auditors' responsibility to check them.

How much checking is an auditor required to carry out? The Companies Act states that the auditors must carry out such investigations as are necessary to form an opinion on the accounts and records, but does not – and indeed could not – specify what that will entail in the case of any particular company. Auditors overcome this problem, or seek to overcome it, by developing methodologies for conducting an audit. Those methodologies, if generally accepted, become the test of what a "reasonable" auditor is expected to do. They specify the checks which he is expected to make and the records which he must inspect. They apply different standards to different types and sizes of company. For example, if a company has a turnover of £100 million, an error of £10,000 in the turnover figure in its accounts will not generally be material. The same error in the accounts of a company with a turnover of £50,000 may be of critical importance.

The auditors' major concern is with the systems which the company has in place. Where a company has established accounting and financial reporting procedures the main job of the auditors is to consider the adequacy of those procedures, establish that they are functioning correctly and carry out random tests on some of the data which they produce. Any deficiencies in the systems will normally be listed in a "management letter" delivered to the board at the end of the audit. This letter has no statutory basis but has evolved as standard auditing practice. If the recommendations in the letter have been adopted by the next audit, that audit should proceed more smoothly. If they have not been adopted by the next audit, the auditors may require an explanation. If there are significant problems with the systems the auditors may raise them in the certificate on the accounts which they produce to the members of the company and may qualify that certificate accordingly.

In the typical company, with the typical problems, questions which arise during the audit can be dealt with adequately in the context of a standard methodology. The difficulty arises with the atypical company or the atypical problem. The problems of auditing a branch of a particular multinational bank have been well rehearsed to date and will no doubt be the subject of comment for years to come. But the context does not have to be exotic for difficulties to arise. Suppose, in the audit of an ordinary trading company, the auditors are unable to obtain satisfactory explanations for the expenditure of amounts which are comparatively small in the overall context of a company's accounts. Suppose further that the explanations of the expenditure which they receive from the directors are not only inadequate but clearly unconvincing. Do they ignore the discrepancy on the ground that the amounts are immaterial in the context of the company's accounts as a whole or insist on investigating it further, thus attracting the hostility of the directors and substantially increasing the fees which the shareholders will be asked to approve? Put shortly, what is the extent of the auditor's duty to investigate possible fraud?

At the moment there is no clear answer. There is indeed what senior members of the accountancy profession refer to as an "expectation gap" between a shareholding public which believes that the role of the auditor is to be a financial detective, tracking every suspicious discrepancy to its source, and a profession which sees its role as to apply a reasonable methodology in a reasonable manner and to publish a note of the results.

While there has been considerable debate on the extent of an auditor's contractual duty there have been comparatively few cases. There are two main reasons for this. One is that professionals such as accountants have a clearer understanding than most of the stress and cost of litigation and tend to settle claims out of court. The other is that there are very few circumstances in which the auditors can have committed an error without the directors also having committed an error. The auditors are, after all, checking the work of the directors. It is, therefore, unlikely that any company would bring a claim against auditors in respect of accounts prepared by the incumbent board: in accusing the auditors, the directors would be accusing themselves. The situation changes when the company is the subject of a takeover. A new board appointed by the purchaser is not implicated in accounts prepared by its predecessors. If those accounts prove to have been inaccurate, and the auditors should have noticed the inaccuracy, the company may well bring a claim against the auditors.

How can we apply these remarks to the new profession of environmental auditors? In fact, the analogies are very close. This is the text of paragraph B of Annex II to the Eco-audit Regulation:

"B. The function of accredited auditors

Validation will be carried out by accredited environmental auditors in conformity with the professional codes of practice which will be established and revised by the Commission.

The function of the auditor is to certify:

- compliance with all the requirements of this Regulation, particularly concerning the environmental review, the environmental audit process and the environmental statements,
- the technical validity of the environmental review or audit process and of any other method and procedure on which the information and data contained in the statement are based,
- the reliability of the data and information in the statement and whether the statement adequately covers all the significant environmental issues of relevance to the site.

The auditor will carry out all the investigations necessary to perform the function described above in a sound professional manner, using appropriate methods and avoiding unnecessary duplication of the company's activities and methods.

The auditor will operate on the basis of a written agreement with the company which defines the scope of the work, enables the auditor to operate in an independent professional manner and commits the company to providing the necessary co-operation.

The verification will involve examination of documentation, a visit to the site including, in particular, interviews with personnel, preparation of a report to the company management and solution of the issues raised by the report.

The documentation to be examined in advance of the site visit will include basic information about the site and activities there, the description of the environmental management system in operation at the site, details of the previous environmental review or audit carried out, the report on that review or audit and the draft environmental statement.

The company will also provide the auditor with the necessary information on previous reviews and audits and details of any infringements of the legislation and of any civil suits relating to environmental matters in the previous five years.

The auditor's report to the company management will specify:

1. cases of non-compliance with the provisions of this Regulation;
2. technical defects in the environmental review or audit method or in any other relevant process;
3. points of disagreement with the review or audit report;
4. points of disagreement with the draft environmental statement, together with details of the amendments or additions that should be made to the environmental statement.

The following cases can arise:

(a) if the protection system, the audit method and the other processes used to produce the statement appear to be adequate and the statement proves accurate, sufficiently detailed and in compliance with the requirements of the Eco-audit scheme, the auditor will validate the statement and certify that it gives a true and fair account of the environmental issues of relevance to the activities at the site;

(b) if the protection system, the audit method and the other relevant processes appear adequate but the statement needs to be revised and/or completed, the auditor will discuss the changes needed with the company management and will not validate the statement until the company has made the appropriate amendments and/or additions to the statement;

(c) if the protection system, the audit method and/or the other relevant processes considered appear inadequate, the auditor will make appropriate recommendations to the company's management on the improvements needed and will not validate the statement until the shortcomings in the processes have been corrected, the processes repeated as far as necessary, and the statement revised accordingly."

All the same elements are included in normal annual accounting requirements:

(i) the primary function is to report on the company's compliance with a statutory (or in this case, non-statutory) obligation, with particular emphasis on the adequacy of records and management systems;
(ii) there is a written agreement with the company which sounds very similar to the financial auditor's letter of appointment;
(iii) there is a "management letter" with recommendations from the auditor which will either be acted on, in which case the environmental statement will be validated (the equivalent of an unqualified certificate on the accounts being issued) or, if they cannot be adequately acted upon, the statement will be revised (the equivalent of a qualified certificate being issued);
(iv) the auditor relies to a large extent on information supplied to him by the company;
(v) the extent of the investigations required of the auditor are specified (initially) in the broadest terms: he must carry out "all the investigations necessary to perform [his] function".

Interestingly, a consideration of this last topic indicates why the English courts have proved reluctant to be more specific as to a financial auditor's duties of investigation. Having given a broad description of the auditor's duty to investigate, the Regulation states: "[t]he verification will involve examination of documents, a visit to the site including, in particular, interviews with personnel, etc." Does that mean that an environmental auditor will be deemed to have carried out all necessary investigations if he only visits the site once?

Liability of auditors to third parties

It is difficult to deal with this topic without rather detailed reference to the relevant decided cases.

A professional man, such as an auditor, owes a duty to the client who employs him "to exercise that standard of skill and care appropriate to his professional status and will be liable both in contract and in tort for all losses which his client may suffer by reason of any breach of that duty" in the words of Lord Bridge in *Caparo Industies Plc* v *Dickman & Ors* [1990] 1 All ER 568 at 575a. He does not generally owe any duty to persons who do not employ him. However, he can owe such a duty if, by his negligence, he causes damage to their person or property – a rather unlikely occurrence for a financial auditor or solicitor but forseeable in the case of, say, an environmental auditor who fails to detect contamination – or if, by a negligent mis-statement, he causes financial loss to such other persons.

The first suggestion that a professional man could, in respect of written or oral statements, owe a duty to a third party with whom he has no contractual relationship arose in *Cann* v *Wilson* (1888) 39 Ch D 39. In that case a person who was intending to borrow on the security of a property employed valuers to produce a valuation of the property. The valuers sent the valuation to the proposed lender's agents and the loan was duly made. When the borrower defaulted and the lender tried to sell the property, it was found that the property had been negligently overvalued and that it was not of sufficient value to cover the loan. The lender recovered his loss from the valuers. Chitty J said: "the document called a valuation was sent by the [valuers] direct to the agents of the [proposed lender] for the purpose of inducing the [proposed lender to make the loan]. It seems to me that the [valuers] knowingly had placed themselves in that position, and in point of law incurred a duty towards [the lender] to use reasonable care in the preparation of the document called a valuation". This is quoted by Lord Bridge in *Caparo* at 575.

However, this proved to be a false dawn for the third party plaintiff because within five years the judgment in *Cann* v *Wilson* had been overruled, in *Le Lievre* v *Gould* [1893] 1 QB 491, and nothing more was heard of the duty of a professional man to third parties until 1951. In that year Lord Denning, in *Candler* v *Crane Christmas & Co* [1951] All ER 426, argued that *Cann* v *Wilson* had been correct and should not have been overruled. The defendants in that case were accountants who had prepared accounts at the request of a company's managing director. Those accounts were shown to a potential investor (the plaintiff) and discussed with him, by the accountants, in the knowledge that he was a potential investor. The accounts were misleading and the plaintiff lost his investment. Lord Denning held that he should be entitled to recover his loss from the accountants. On that occasion Lord Denning failed to persuade his colleagues and the plaintiff failed. However, judicial opinion was moving in his direction and in *Hedley Byrne & Co Ltd* v *Heller Partners Ltd* [1963] 2 All ER 575, the House of Lords adopted Lord Denning's minority judgment in *Candler* v *Crane Christmas*, overruled the majority judgment in that case and reinstated *Cann* v *Wilson*.

The *Hedley Byrne* case concerned bankers who had negligently given a reference on behalf of a customer on the faith of which the plaintiff gave credit to that customer. The customer defaulted and the plaintiff sued the bank. The court held that the plaintiff was entitled to recover from the bank – or rather that he would have been entitled to recover if the bank had not inserted a disclaimer of liability in the reference. It is important to note that the bank did not know the identity of the lender when it gave the reference although it must necessarily have concluded that the purpose of the request for a reference was to assess the customer's credit-worthiness. This factor distinguishes the case from *Cann* v *Wilson* and *Candler* v *Crane Christmas* where the identity of the third party plaintiff was known to the defendant at the time the duty arose.

At this point, the law of *negligent mis-statement*, as the rule in *Hedley Byrne* was known, came under the influence of the more general law of negligence. The law of negligence had initially proceeded on the basis of precedent and analogy. In considering a case the court scrutinised the relevant set of facts to establish to what extent there were analogies with the facts in previous cases where a "duty of care" had been held to have arisen. Only if the analogies were close would a duty of care be found. There was a movement among certain judges, however, to put the law of negligence onto a more "codified" basis. They scrutinised the preceding cases in an attempt to adduce general "tests" which could be applied more readily to cases in the future. Their aim, express or implied, was to reach a position in which it would not be necessary to show that the facts of a new case were analogous to those of an existing case, but only to show that the facts of that case satisfied general tests for negligence. It was, perhaps, inevitable that a consequence of this approach was to extend the boundaries of negligence, although the proponents of the approach proposed to limit that extension by defining "exceptional" circumstances in which, although the duty of care arose, it would not (usually for reasons of public policy) be actionable. The culmination of this enterprise was the leading judgment of Lord Wilberforce in the House of Lords case, *Anns* v *Merton London Borough* [1978] AC 241. This judgment influenced a New Zealand decision on negligent mis-statement, *Scott Group Ltd* v *McFarlane* [1978] 1 NZLR 553. The case is now notable mainly for the fact that Richmond J, in a dissenting judgment approved in *Caparo*, first used the phrase "special relationship" to describe the relationship which must subsist between the adviser and the third party.

In an English case, *JEB Fasteners Ltd* v *Marks Bloom & Co (a firm)* [1981] 3 All ER 289 and [1983] 1 All ER 583 (CA), Mr Justice Woolf dismissed a claim against auditors on the grounds that the plaintiff had failed to show any "causative connection" between reliance on the erroneous accounts and a subsequent takeover and his judgment was upheld by the Court of Appeal.

However, in the course of the judgment he made various additional statements about forseeability which were criticised in *Caparo*. This and the Scottish case of *Twomax Ltd* v *Dickson McFarlane & Robinson* 1982 SC 113 were cases where the general rules of forseeability referred to in Lord Wilberforce's judgment were applied to auditors who had negligently audited a company's accounts.

Lord Wilberforce's approach did not attract unanimous support. Subsequent cases on negligence, *Peabody's Donation Fund* v *Sir Lindsay Parkinson & Co Ltd* [1984] 3 All ER 529, [1985] AC 210 and *Yuen Kun-yeu* v *A-G of Hong Kong* [1987] 2 All ER 705, [1988] AC 175, and *Sutherland Shire Council* v *Heyman* (1985) 60 ALR 1, were quoted or referred to with approval by all the judges in *Caparo* (*e.g.* Lord Bridge at 574d). Interestingly, the other classic statement of this approach is in Lord Devlin's judgment in *Hedley Byrne* itself. It is quoted at length by Lord Oliver in *Caparo* at 586d to 587a. The Law Lords found that his statements in *Anns* had been "misinterpreted". There were no general tests. Indeed the attempt to find a satisfactory general test was doomed to failure. The reaction is best expressed in a judgment of an Australian judge, Mr Justice Brennan:

> "It is preferable in my view, that the law should develop novel categories of negligence incrementally and by analogy with established categories, rather than by a massive extension of a prima facie duty of care restrained only by indefinable 'considerations which ought to negative, or to reduce or limit the scope of the duty or the class of person to whom it is owed'."

It was in that context that three cases on negligent mis-statement came for consideration before the House of Lords in 1989. Two of them, *Smith* v *Eric S Bush (a firm)* and *Harris and anor* v *Wyre Forest District Council & anor* concerned surveyor valuers. They were on very similar facts and were heard together. The third, *Caparo Industries plc* v *Dickman & Ors*, concerned auditors.

The *Smith* and *Harris* cases were summarised by Lord Bridge in *Caparo:*

> "The plaintiffs in both cases were house purchasers who purchased in reliance on valuation of the properties made by surveyors acting for and on the instructions of the mortgagees proposing to advance money to the plaintiffs to enable them to effect their purchases. In both cases the surveyors' fees were paid by the plaintiffs and in both cases it turned out that the inspections and valuations had been negligently carried out and that the property was seriously defective so that the plaintiffs suffered financial loss. In Smith's case the mortgagees were a building society, the surveyors who carried out the inspection and valuation were a firm employed by the building society, and their report was shown to the plaintiff. In Harris's case the mortgagees were the local authority who employed a member of their own staff to carry out the inspection and valuation. His report was not shown to the plaintiff, but the plaintiff rightly assumed from the local authority's offer of a mortgage loan that the property had been professionally valued as worth at least the amount of the loan . . . The House held that in both cases the surveyor making the inspection and valuation owed a duty of care to the plaintiff house purchaser . . .."

Caparo concerned auditors. Fidelity plc owned shares in Caparo Industries plc.

When Fidelity published its accounts for the year ending 31 March 1984, the reported profits were far short of what had been anticipated and the share price fell dramatically. Having received the accounts, Caparo immediately purchased more shares in the company and later that year made a successful takeover bid for the company. Some time after the purchase, Caparo claimed:

(i) that the March 1984 accounts had been inaccurate in that the reported profit of £1.2 million should in fact have been a loss of £400,000;
(ii) that the auditors had been negligent in auditing the accounts;
(iii) that Caparo had purchased further shares and made its takeover bid in reliance on the accounts; and
(iv) that the auditors owed them a duty of care either as potential bidders for Fidelity (because they ought to have seen that the 1984 results made Fidelity vulnerable to a takeover bid) or as an existing shareholder in Fidelity interested in buying more shares.

The trial concerned the preliminary issue of whether the auditors owed a duty of care to Caparo.

It follows from the comments on the reaction against *Anns* that the various judgments in the House of Lords proceeded, not on the basis of general principles, but by analogy with previous cases, although both Lord Oliver and the draftsman of the headnote to the All England Report seemed reluctant to abandon the terminology of "forseeability" and "proximity".

In the passage immediately following that quoted above, Lord Bridge continued:

> "The salient feature of all these cases [*i.e Cann* v *Wilson, Hedley Byrne, Smith* and *Harris*] is that the defendant giving advice or information was fully aware of the nature of the transaction which the plaintiff had in contemplation, knew that the advice or information would be communicated to him directly or indirectly and knew that it was very likely that the plaintiff would rely on that advice or information in deciding whether or not to engage in the transaction in contemplation. In these circumstances the defendant could clearly be expected, subject always to the effect of any disclaimer of responsibility, specifically to anticipate that the plaintiff would rely on the advice or information given by the defendant for the very purpose for which he did in the event rely on it. The situation is entirely different where a statement is put into more or less general circulation and may forseeably be relied on by strangers to the maker of the statement for any one of a variety of different purposes which the maker of the statement has no specific reason to anticipate. To hold the maker of the statement to be under a duty of care in respect of the accuracy of the statement to all and sundry for any purpose for which they may choose to rely on it is not only to subject him, in the classic words of Cardozo CJ to 'liability in an indeterminate amount for an indeterminate time to an indeterminate class (see *Ultramares Corp* v *Touche* (1931) 255 NY 170 at 179), it is also to confer on the world at large a quite unwarranted entitlement to appropriate for their own purposes the benefit of the expert knowledge or professional expertise attributed to the maker of the statement'."

Lord Bridge went on to refer to:

> "the necessity to prove, in this category of the tort of negligence, as an essential ingredient of the 'proximity' between the plaintiff and the defendant, that the defendant knew that his statement would be communicated to the plaintiff, either as an individual or as a member of an identifiable class, specifically in connection with a particular transaction or transactions of a particular kind (*e.g.* in a prospectus inviting investment) and that the plaintiff would be very likely to rely on it for the purpose of deciding whether or not to enter on that transaction or on a transaction of that kind."

In *Caparo* the auditors did not know that Caparo had any intention to purchase shares although they clearly knew that the accounts would be sent to all existing shareholders of the company. Both the Court of Appeal and the House of Lords found it relatively easy to dismiss the suggestion that there was some identifiable class of potential purchasers of companies in difficulties whom auditors would always have in contemplation in auditing the accounts of such companies. This overturned *Twomax Ltd* v *Dickson McFarlane & Robinson* in the process. The more difficult question was whether because of the statutory duties imposed on auditors – duties which, it was argued, had as a primary object the protection of shareholders – there was to be implied a "special relationship" between the auditors and shareholders which was analogous to the "special relationship" found in *Cann* v *Wilson*, Lord Denning's dissenting judgment in *Candler* v *Crane Christmas*, *Hedley Byrne*, *Smith* and *Harris*. A majority in the Court of Appeal found that such a special relationship existed: the House of Lords preferred the minority judgment of Lord Justice O'Connor and unanimously found that there was no special relationship.

The grounds on which the Lords reached their decision are best set out in Lord Oliver's judgment. Like Lord Bridge he laid stress upon the facts:

(i) that in the earlier cases the defendants had known the transaction, or the general nature of the transaction, in which the plaintiff intended to participate; and
(ii) that they had also known that the negligent mis-statement in question would probably (not *possibly*) be relied upon by the plaintiff for the purpose of that transaction.

It was argued by Caparo that statutory accounts were prepared for the purpose of enabling shareholders "to make such individual decisions as they wished in relation to the company, including decisions as to investments, they already being investors, and decisions as to voting in general meetings." It followed that auditors delivered accounts to shareholders in the knowledge that they would probably rely on those accounts for the purpose of deciding whether to buy or sell shares in the company. Therefore, although the auditors in this case did not know that Caparo intended to purchase shares, Caparo fell within an identifiable class of people whom they knew might contemplate purchasing shares (*i.e* existing shareholders) and who would probably rely on the accounts in doing so.

Lord Oliver disagreed. He found that:

> "the purpose of annual accounts, so far as members are concerned, is to enable them to question the past management of the company, to exercise their voting

> rights, if so advised, and to influence future policy and management. Advice to individual shareholders in relation to present or future investment in the company is no part of the statutory purpose of the preparation and distribution of accounts . . . If the statutory accounts are prepared and distributed for certain limited purposes, can there nevertheless be imposed upon auditors an additional common law duty to individual shareholders who choose to use them for another purpose without the prior knowledge of the auditors? The answer must be No. Use for that other purpose would no longer be use for the 'very transaction' which Denning LJ in Candler's case . . . regarded as determinative of the scope of any duty of care. Only where the auditor was aware that the individual shareholder was likely to rely on the accounts for a particular purpose such as his present or future investment in or lending to the company would a duty of care arise. Such a situation does not obtain in the present case."

Caparo has been followed by a flurry of cases in which the limits of the decision have been tested in the hope of finding a situation in which a "special relationship" would arise.

One, which was decided after the Court of Appeal judgment in *Caparo* but before the House of Lords judgment, was *Huxford & ors* v *Stoy Hayward & Co*, *The Times*, 11 January 1989. It is slightly off the point in that it concerned an attempt by individuals, who were directors and guarantors of, and shareholders in, a company, to argue that accountants, brought in at the instigation of the bank to advise the directors on the financial position of the company, also owed a duty of care to the directors in their capacity as guarantors and shareholders. They failed to convince Mr Justice Popplewell.

More interesting is the case of *Morgan Crucible Co* v *Hill Samuel Bank Ltd & ors* Ch D, Mr Justice Hoffman, The *Independent* 27 July 1990. Morgan Crucible made a formal bid for a company. The chairman and advisers of that company sent circulars to shareholders recommending that the bid was refused. The circulars, which were also issued as press releases, made various statements about the future profits of the company. In response to the circulars, Morgan Crucible raised its bid and finally succeeded in acquiring the company's shares. It later claimed that the profit forecasts had been made negligently and that the chairman, directors and advisers (the merchant bank and auditors) had owed Morgan Crucible a duty of care. It argued that, in contrast with the situation in *Caparo*, the advisers knew that Morgan Crucible were intending to purchase shares and they also knew that Morgan Crucible were likely to rely on the circulars in deciding whether or not to increase their bid. They admitted that the circulars were not prepared for Morgan Crucible but for the company's shareholders but claimed that that put the case "on all fours" with the *Smith* case where the valuation had expressly said it was intended for the building society and should not be relied on by the purchaser.

Hoffman J distinguished *Smith* from *Caparo* on a number of grounds, most of which could be described as socio-economic. Mr Smith had paid for the survey; Caparo had not paid for the audit. Mr Smith was a person of modest means making the most expensive purchase of his life; a takeover bidder, on the other hand, was an entrepreneur taking high risks for high rewards. A surveyor could probably cover himself against claims of the *Smith* type by insurance;

some accountants might be unable to take out sufficient insurance to cover themselves against claims of the *Caparo* type. The cost of such insurance to a surveyor would be comparatively low and would not cause any significant economic effect: "take-overs on the Stock Exchange involve huge amounts and the effects on accountants' insurance and fees were unpredictable."

Against that background he found that Morgan Crucible's position was much closer to that of Caparo than that of Mr Smith. In addition he stated that the defence documents were prepared pursuant to an obligation imposed by the Takeover Code which required the directors to advise the shareholders whether or not to accept a bid. The accountants might very well owe a duty to the individual shareholders:

> "to take reasonable care to ensure that they did not sell their shares too cheaply. But that had nothing to do with whether they owed a duty to the bidder to take care that he did not pay too much. The case could not be distinguished from *Caparo*. There was a sense in which the reasons for denying the duty of care were even stronger than in *Caparo*. In a contested bid, the interests of the bidder and the company's shareholders were in conflict with each other. The board was under a positive duty under the Code to advise its shareholders. To impose on the board and its advisers a duty of care owed to the bidder would put them in a difficult position. It could be done but the position was nevertheless uncomfortable and undesirable. The reality of a contested take-over bid was that you had to know what side you were on."

In conclusion, liability of auditors to third parties will generally be difficult to establish. The law of negligence has undergone substantial retrenchment since the days of Lord Denning. However, just as environmental auditors may start to breath a huge sigh of relief, architects and engineers in the construction industry enjoying similar immunity saw the arrival and subsequent permanent landing of contractual warranties and indemnities.

CONTENTS OF CHAPTER 9

Environmental Risk Protection: Warranties, Indemnities, Insurance and Confidentiality

Chapter 9
Environmental Risk Protection: Warranties, Indemnities, Insurance and Confidentiality

Much of this book has been concerned with the move towards environmental standards and the development of Eco-audits. Rather than embark upon enforcement of laws alone, the European Commission, in particular, is moving towards encouraging improvements in environmental performance by the implementation of proper and effective environmental management systems. This leads to an increase in workload and responsibility and results in the production of documents. This chapter looks at ways of off-setting some of the risks which the increasing volume of legislation is creating through warranties, indemnities and insurance, and at ways of seeking to maintain confidentiality in respect of necessary documentation.

Warranties and indemnities are a sensible legal device but their effectiveness is subject in particular, as experiences in the construction industry show, to the risk of insolvency. To mitigate this difficulty, environmental insurance has developed, where, hopefully, the regulations governing insurance companies and their assets make insolvency less of a risk. But, just as legislation may require disclosure of information, so too may both warranties and insurance. In this indirect way, issues of confidentiality may arise.

Warranties and indemnities

Chapter 3 looked at the impact that the adoption of an environmental management policy may have upon a company's relations with its shareholders, suppliers and customers and with the regulatory authorities. Chapter 7 considered the effect of requirements for impact assessments before land development can proceed, and the impact of those assessments and contamination on land values and liabilities.

There is, however, another situation in which the risk of liability associated with environmental damage or non-compliance with environmental legislation has taken on particular importance over the last few years. This is the sale of businesses or shares. When a purchaser is buying a company or, increasingly, a business, it has become common for him to ask for certain warranties from the vendor. "Warranty" is unfortunately a word which is used in several different contexts by lawyers. In relation to the sale of a company, it means a series of statements which the vendor guarantees to be true, knowing that the purchaser is relying on their accuracy in agreeing to purchase the company or business in question for the agreed price. Some of the so-called warranties would be implied at law in any event – for example, that the vendor is the true owner of the shares which he is purporting to sell – but warranties of the type we are discussing here do not fall into this category. They cover matters which, if not expressly dealt with in the agreement, would pass to the purchaser as liabilities or potential liabilities on the purchase of the company. For example, the purchaser of shares in the company will normally ask for a warranty that corporation tax is paid up to date and that there is no outstanding dispute with the Inland Revenue concerning past assessments. If he does not ask for such a warranty, and the company turns out to owe £10 million in back-taxes, then he would, in the normal course of events, have no remedy against the selling shareholder. If the general rule is *caveat emptor*, then warranties are the exceptions to the general rule.

Like a surprising number of areas of the law and legal practice, warranties are subject to fashion. In the early 1980s there was concern about asbestos and other deleterious materials being used in the fabric of buildings, and the purchaser of a company would seek a multitude of warranties to the effect that various materials had not been used in the construction of any of the company's buildings. Typically there would, however, be no warranties as to whether the buildings in question were constructed on contaminated land. In the early 1990s the presence of deleterious materials had not proved as significant as at first thought and the scope of such warranties generally has become more restrictive and less onerous.

In the absence of any express warranty, the risk of contamination is one which is assumed by the purchaser. That is not necessarily unreasonable. There are risks that are inherent in every business and it is no more unreasonable to suggest that the purchaser should take the risk of unsuspected contamination than that he should take the risk of unsuspected insolvency of one of the company's major debtors. The distinction that is drawn between such warranties is that a warranty of the land relates to events occurring in the past, which are capable of investigation by the vendor, whereas a warranty on the collectability of debts extends to events occurring in the future which are not within the control of the vendor. That distinction is unconvincing, however, where the events occurring in the past were neither under the control of the vendor nor capable of investigation without unreasonable expense.

The vendor who is asked in 1992 to warrant that none of the company's properties is situated on land which is, or ever has been, contaminated is in very much the situation of the vendor who was asked in 1982 to warrant that none of the company's properties had been constructed using any deleterious materials. It is unlikely that he will know the answer to the question and he will share the same reluctance to commission a full environmental survey of all sites which his predecessor in 1982 would have shown to have samples taken of all building materials used in his buildings. If the purchaser persists with his request for reassurance on this aspect of the company's affairs, the vendor has four choices. He can:

(i) refuse to give the warranty and risk losing the sale;
(ii) commission an environmental audit for his own benefit so that he can assess what risks there might be in giving the warranty;
(iii) refuse to give the warranty but allow the purchaser to carry out an environmental audit (thus risking losing the sale if the purchaser's fears are confirmed); or
(iv) seek to modify the warranty in some way in order to balance the risks to the purchaser and the vendor.

For example, if the vendor had kept records of the use of all the sites since, say 1950, he might be prepared to warrant that there had been no contamination of the site since that period (save for specific incidents of which he was aware) but refuse to give any warranties on previous use. That might not be sufficient for the purchaser. He might want, as a minimum, to know that the vendor "does not, having made diligent and comprehensive enquiries, have any reason to believe that the land was contaminated before 1950." This raises the question not only of the meaning of "contamination"' but of what enquiries could be considered to be diligent and comprehensive in this context.

Even in this situation, however, a compromise may be possible. The purchaser requires a warranty that the factory site is not contaminated. The vendor refuses to give it. One of the grounds on which the vendor bases his refusal is the fact that, if the land proved to be contaminated in the future, the vendor will be unable to control the remedial measures taken to deal with the contamination and, in particular, the cost of those remedial measures. He might also fear that the remedial measures might lead to consequential loss because of a temporary or permanent closure of the factory while the contamination was removed. Those fears can be met to some extent by restricting the amount of the vendor's liability either generally or by specific reference to a claim under the warranty in question. If the purchaser was only able to bring a warranty claim within the first three years after acquisition then the vendor would effectively be taking the risk only of contamination which became apparent during that period. Similarly his financial liability could be restricted to a specified maximum sum.

That still leaves the matter rather uncertain. The vendor cannot be sure that the purchaser will not dig up the site in order to discover whether there is contamination in time to bring a claim under the warranties. The purchaser is faced with the possibility that contamination will emerge after the warranty period has expired or that the cost of remedial works will far exceed the maximum limit on any claim. Where there is a specific problem of this nature which is of considerable concern to both vendor and purchaser, an indemnity or specific financial provision can often be a means of breaking a deadlock.

For example, the vendor and purchaser might agree that after the acquisition a specified number of core samples would be taken at points on the site specified by the purchaser or its advisers. Those samples would be analysed and, if they were found to contain contamination, the vendor would be required either to pay a fixed sum in compensation to the purchaser or to carry out remedial works at its own expense. If the factory were critical to the operation it could even be possible that the purchaser had a right to rescind the contract and require the vendor to take the company back. The corollary of this provision, however, would be that if the core samples showed no contamination, the purchaser would not be entitled to bring any claim against the vendor in respect of the site even if subsequent events should prove that unassembled areas of the site were, indeed, subject to contamination. A provision such as this would have to be carefully drawn so that both sides were satisfied as to the impartiality of the persons testing the site and were agreed as to the definition of "contamination" for this purpose. Different figures could be payable depending on different degrees of contamination. At the end of the day, it is a question of balance of risk – a question, ultimately, of whether the purchaser's enthusiasm to buy the company exceeds the vendor's enthusiasm to sell it.

The effect of an existing environmental management policy

Would this balance of risk between vendor and purchaser be affected by the fact that the company in question had adopted an environmental management policy?

In many ways, the existence of an environmental management policy should make the situation simpler. The vendor asked to give warranties as to his compliance with environmental legislation, the existence of contaminated land, etc, can be expected to deliver to the purchaser a copy of the latest environmental statement for each site and the accompanying auditors' verification and ask him to rely on those documents to the exclusion of any more general warranty. This solution would be analogous to the solution normally adopted in the case of audited financial accounts. A purchaser, purchasing close to the year end of a company, will normally be prepared to rely on a warranty that audited statutory accounts have been prepared in accordance with all relevant legislation and accounting standards and in accordance with general accepted accounting principles (subject, of course, to any comments in the notes to the accounts or the auditors' certificate showing a divergence from such principles). It is only when a considerable period has elapsed since the year end or where there are particular items of concern that more detailed accounting warranties will be sought.

The type of warranties which might, therefore, be offered and accepted would cover the following matters:

(i) that the environmental management policy of the company was as set out in certain specified copy documents;
(ii) that the environmental management system for each site was contained in further specified documents;
(iii) that the last environmental statement issued by the company was as set out in specified disclosed documents and that the period stated in those statements for the production of the next environmental statement had not yet expired;

(iv) that the last auditors' certificate for each site was in the form of a specified document;
(v) that all recommendations contained in the last auditors' certificate had been implemented – or that they had been implemented to the extent specifically disclosed in the disclosure letter;
(vi) that those of the objectives contained in the last environmental statement which were to be complied with before the issue of the next statement either have been complied with or will be complied with within the specified period (without undue difficulty or expense); and
(vii) that the records of the company since the date of the last environmental audit do not show any notifiable incidents other than those disclosed in the disclosure letter (notifiable incidents in this context might be incidents that were required by the environmental management system, to be notified and recorded in the company records in accordance with the company's management policy).

From the vendor's point of view there are certain advantages in giving warranties in this form. Most important, is that he can check whether each of the warranty statements is true. From the point of view of a purchaser, he can take some comfort from the fact that the company has investigated its potential environmental liability and that its investigations have been considered by an outside auditor. There may, however, be areas where the purchaser still has reasonable grounds for concern. Primarily, the fact that a system exists for detecting the contamination of a site does not mean that the site is not contaminated. The system might be inadequate to detect the particular form of contamination involved, or it might have been applied negligently or it may simply not have been applied at all.

Take, for example, a factory built close to a city centre in 1950. Since 1950 the factory has not engaged in any processes which could contaminate the land and prior to the Second World War the land on which it is built was used for housing. The housing, however, was destroyed by bombing in the war and the likelihood is that there were bomb craters in the land on which the factory was erected. Those bomb craters would have been filled up either before, or in the course of, the building of the factory. It is possible, therefore, that contaminated materials were used as part of the land-fill.

As part of its initial environmental review, the company has discovered all the foregoing information. It receives advice that the only means of determining whether any of the land-fill material is contaminated is to take a series of core samples at various points on the site. These samples will require holes to be drilled in the floor of the factory and will not necessarily be conclusive if a more-or-less random mixture of materials has been used for the in-fill – it is always possible that the bore holes will miss a pocket of contaminated material.

It is no necessary consequence of a responsible environmental policy that the company proceed to drill the bore holes. The company might well decide that, given the cost and disruption caused by the investigation and the limited prospect of a conclusive result, the risk of contamination is not sufficient to justify further investigation. The position of the directors would be similar to the position of directors who, when preparing a company's statutory accounts, decide that the chances of a particular contingent liability being realised are sufficiently remote for no provision to be made or noted in the company's accounts. In both cases, external auditors will review the directors' decision but provided that they have no ground on which to consider the directors' assessment of the risk unreasonable, they are unlikely to subject it to adverse criticism. A purchaser should, therefore, not assume that the existence of a warranted environmental management policy will allow him to make a claim.

Balancing the risk

The final terms of warranties and indemnities depend upon a number of factors including the bargaining position of the parties and the importance and/or significance and/or sensitivity of the subject matter to which the warranty applies. There are examples of vendors asking for strict liability warranties *i.e.* that all environmental legislation has been complied with, or alternatively seeking such warranties from those who have, on behalf of the vendor, undertaken environmental audits that they themselves have in the execution of their work complied with all environmental legislation or, perhaps to a lesser extent, have exercised reasonable skill and care in the execution of their works.

The value of warranties from third parties (*e.g.* environmental consultants) is that they give a prospective vendor an additional – usually insured – contractual right to recover damages from them. In the absence of legal safeguards, the vendor may have two parties against which to make a claim, while an environmental auditor inadvertently gives two rights of recovery for the one fee.

The value of warranties can be seen in the construction industry where 15 years ago warranties were not in demand. The growing re-trenchment of the law of negligence limiting the scope and extent of the duty of care witnessed the expanding demand and eventual acceptance by the industry that warranties were here to stay. One would anticipate that on environmental issues, with the increase in Eco-audits, that warranties and indemnities will follow a similar path and are likely to become widely used over the next few years.

At the end of the day it remains a question of identifying and balancing risk. Neither is easy and in circumstances where one party may feel uneasy in his assessment of risk, there remains the option of insurance – the ability of transferring a risk or part of it to an insurance company in return for a premium. This in itself is a complicated and, at times, a mysterious path to travel, particularly in relation to environmental issues and such problems as the gradual process of pollution. Before embarking upon such a journey, one needs to begin at the beginning and reflect, albeit briefly, on the general principles of insurance law.

Insurance - a legal introduction

Insurance policies tend to be rather complicated because of the way they are set out. They will recite clauses giving details of their coverage. There will then be a section which deals with exceptions, *i.e.* events that are not covered, and there will then be other parts which amount to exceptions to exceptions – often an event that was specifically excluded in the first exception can actually come back into cover by the exception to the exception. One should also say that the real difficulty with insurance is that the insurance market often appears to regard the extent of cover as being determined by their understanding as to what the cover is, rather than by the words used: there are many examples of cases where policies which are issued do in fact give wider cover, or in certain circumstances give narrower cover than the insurer believes. This, needless to say, often leads to litigation arising between the insured and the insurer where the insured is claiming more than perhaps the insurer believed his policy covered. One of the best examples in the last few years is that of building contractors who had during the early 1980's managed to obtain insurance cover on their duty to supervise. As policies were renewed annually, insurers had sought to limit this type of cover.

In this context, particular regard should be had to five different types of insurance policies that are generally available on the market. Later on in this chapter there are specific references to policies that are available. However, it is for each potential insured to look carefully at the words and the terms of the policy, the amount of the premium and the excess involved. The types most commonly encountered are:

- Contractors All Risk
- Public Liability (and similarly Product Liability)
- Professional Indemnity Insurance
- Environmental Impairment Liability (EIL) Insurance
- Directors and Officers Liability Insurance (personal insurance discussed later in this chapter).

The Contractors All Risk policy, as its name implies, gives a contractor the widest possible cover against damage to the works. Its object, as the operative clause provides, is to indemnify the insured for loss, damage or destruction by any cause whatsoever of the contract works and includes the materials and goods on site and ought to cover temporary works, tools, plant, equipment and temporary buildings etc. However, insurers have started to add an exclusion clause omitting debris, removal and disposal costs or expenses arising from pollution or contamination of property not insured by the policy. An example of this is a recent amendment by Sun Alliance International.

Public Liability policies provide an indemnity against legal liability for bodily injury and property damage claims by third parties. Generally speaking, Public Liability policies would not cover professional negligence, and in the United Kingdom these policies exclude claims for gradual pollution. Product liability insurance is similar to public liability insurance but is issued in relation to the potential for products to cause damage to third party property or to third persons.

The third type of policy is the Professional Indemnity policy which normally protects the insured against his legal liability to pay damages to third parties who have sustained some injury, loss or damage due to the negligence of the insured or, of course, his staff in the course of the professional conduct of its business. Generally speaking liability is based on a breach of a professional duty which means a failure to exercise the ordinary degree of skill and care expected of a competent practitioner engaged in the profession concerned.

A professional man's position in law can best be described by reference to the words of Tindal CJ in *Lanphier* v *Phipos* (1838):

> "Every person who enters into a learned profession undertakes to bring to the exercise of it a reasonable degree of care and skill. He does not undertake, if he is an attorney, that in all events you shall gain your case, nor does a surgeon undertake that he will perform a cure, nor does he undertake to use the highest possible degree of skill. There may be persons who have a higher education and greater advantages than he has but he undertakes to bring a fair, reasonable and competent degree of skill" (*i.e.* to exercise the ordinary skill and care expected of a competent practitioner engaged in the profession concerned.)

As environmental auditors seek to obtain professional indemnity cover, an interesting development will be clarification of what is meant by the ordinary skill and care expected of such an environmental auditor. The level of knowledge which an auditor should display, and how he should approach his task, is likely to rise with the development of the Eco-audit, and BS 7750 (and its European equivalent) as these standards are likely to be taken as indicating the kind of level of care that is required.

The other point to bear in mind is that professional indemnity insurance arises out of exercising reasonable skill and care and tends not to bite in circumstances where it is a form of no-fault obligation which is giving rise to the liability. This is particularly pertinent in relation to European Directives which, following the approach of many continental, civil law systems, tend to be based upon strict liability. As such, parties who have agreed to enter into no-fault liability obligations either in contract or through direct compliance with European Directives may find that their professional indemnity does not give cover from potential environmental risks.

Environmental Impairment Liability insurance normally covers the insured's legal liability for third party injury or property damage arising out of an environmental occurrence such as the discharge, disbursement, release or escape of smoke, vapours, fumes, acids, alkaline, toxic chemicals, liquid or gases, waste materials or other irritants, contaminants or pollutants into or upon land etc, *i.e.* pollution events. These policies are normally let on a claims made basis and more detail on this and the cover that is now available is discussed later in this chapter.

Finally, there is Directors and Officers (D&O) liability insurance. With the increasing range and level of fines which are a feature of the environmental enforcement scene, it might be hoped that D&O policies would offer some source of comfort for those in positions of responsibility in administering health and safety matters. Unfortunately for them, liability arising in connection with pollution is now generally excluded.

General principles of insurance

As a matter of English law (insurance law may vary from one EC Member State to another) insurance arrangements are made by contract, and the usual rules as to the formation of contracts by means of an offer and acceptance and consideration apply to contracts of insurance in the same way as they do to any other contracts. In other words there is, in basic terms, no difference between having a contract for the supply of goods and contracts in relation to the supply of insurance cover: an offer is made to buy the goods and the goods are provided for a sum of money. The normal procedure is for a potential insured to complete a proposal form, usually the insurer's standard form, which will refer to the standard conditions of the policy. By signing the proposal form the customer is offering to accept insurance on those terms. The terms themselves usually contain a provision that the proposal form is incorporated into and is the basis of the insurance contract. In this way all the information given by the customer is in the proposal form which becomes part of the insurance contract. It follows that once the proposal form has been given to the insurers and the insurers have accepted it, the contract will exist and the insurers are obliged to issue a policy in the terms agreed between the parties.

Although the rules as to formation of a contract are the same, an insurance contract has certain rules and principles in addition to those normally applied in the law of contract. While good faith must, of course, be observed in all contracts, at least in the sense that the parties must not act fraudulently, this does not generally mean that they are bound to reveal all they know or ought to know about a particular transaction. Indeed, many arrangements depend for their profitability on restrictive access to information. Contracts of insurance are, however, quite different because only one party to the contract, the proposer, knows or ought to know all about the risk proposed for insurance and the other party, the insurer, has to rely mainly upon information given to him by the proposer in his assessment of that risk. For this reason contracts of insurance are described as contracts *uberrimae fidei*, which means "of the utmost good faith". By way of example the duty of utmost good faith is not discharged by an insured simply by replying truthfully to all the questions on a proposal form. If there are any material facts which might affect the underwriters minds in considering the risk, they must be disclosed.

The principle of good faith applies to both parties and where utmost good faith has not been shown by one party, the other party can avoid the insurance contract. It is of course vital that the proposal form, which usually forms the basis of the contract, is completed properly and great care must be taken in ensuring that it contains:

(i) disclosure of all material facts; and
(ii) no statements that would amount to a misrepresentation of material facts.

In addition to disclosing all material facts that are actually within his knowledge the proposer cannot necessarily assume that the matters that are not within his knowledge are not disclosable. Rather, the insured must disclose to the insurer all facts material to an insurer's appraisal of the risk which are known or deemed to be known to the insured. The proper question in each case is whether a particular circumstance was in fact material not whether the proposer believes it to be so.

The person seeking to insure may fairly be presumed to know all the circumstances which materially affect the risk, and, generally, is as to some of them the only person who has knowledge. By way of comparison, the underwriter whom he asks to take the risk cannot, as a rule, know and rarely has either the time or the opportunity to learn by enquiry, circumstances which may be material to the formation of his judgment in deciding whether to accept or reject the risk.

By way of summary it is incumbent on the insured to disclose all facts which a reasonably prudent underwriter would take on board when assessing the risk. The question of reasonableness is not judged by the court or by the insured, it is a test which is judged by what a prudent underwriter would take on board by way of assessing the risk. For example, with regard to Professional Indemnity insurance material facts would include facts that indicate the previous experience of a professional man in relation to matters that are relevant to the insured such as any previous relevant claims against the insured for professional negligence. Another material fact would be a particular area of activity in which the professional person operates since clearly some professional operations are far more hazardous than others from an insurance point of view. An example is the giving of advice in the United States because, compared to advice given in Europe, there is statistically a much greater chance of proceedings being taken in the event that the advice is regarded as being unsatisfactory.

There is also a duty to disclose material facts not only at the time of completing the proposal form but also any further material facts which come to the insured's knowledge before the contract of insurance is concluded.

Non-disclosure

A policy will be avoided *ab initio* for non-disclosure. This is where a material fact has not been disclosed. The effect is that the insured becomes uninsured in respect of all claims subject to the policy. Although he is entitled to recover the premiums that he has paid, insurers are entitled to recover previous claims paid out under the policy.

In contrast, the effect of late notification (see below) is normally to avoid the claim but not to avoid the policy. An area where a proposer ought to be careful is when he changes his insurers. For instance, it may be on the last day of his existing policy that he notifies his present insurers of a claim. However, if he takes out a new policy with new insurers for the following year without notifying them of his previous claim under an old policy, that in itself could be a material fact entitling insurers to avoid the policy by reason of non-disclosure.

Notification and claims

The policy will usually contain a condition requiring the insured to give written notice to the insurer of claims made against the insured. The precise wording of this condition is of fundamental importance because observance of the procedure under a policy for giving notice to insurers is usually stated to be a condition precedent to the insurer's liability. The insured should therefore read the conditions setting out the procedure for notification on claims with the greatest care so as to be certain that he will be able to comply with it.

The usual wording requires the written notice to be given when the insured becomes aware of circumstances which may give rise to a claim. Such wording means that the insured cannot wait until the claim is actually made against him: rather, as soon as he knows the circumstances which may lead to such a claim, he has a duty to inform insurers. If the condition requires the notification to be in writing, then it must be in writing. This also helps to avoid any argument as to the precise date when notification was made. The other difficulty relates to the level of notification that should be given. By way of example, the dilemma is that if, as an architect, an insured is worried about possible design defects, every time a potential problem arises on any development then, on one view, he ought to notify insurers of possible claims even if he reasonably considers that no liability will arise. The problem is that erring on the side of safety and continuing to notify in this way may have an impact when renewing the policy; *i.e.* there may be an increase in premiums. Equally, if an insured takes a more robust view that many of the day-to-day complaints received are not likely to amount to any claim and he, therefore, does not give notification in respect of them, he runs the risk of notification difficulties. It is, at the end of the day, a matter that needs to be discussed and reviewed between the insured and the insurer, or as is more likely, with the insurance broker. It is not easy to be more specific because it is a matter of judgment as to when to notify. However, as a general guide, it is better to err on the side of caution and, if in doubt, to notify.

Insurers normally retain rights of subrogation in the event of a claim. Subrogation is in effect where insurers stand in the shoes of the insured and avail themselves of all the rights and remedies of the insured but obviously only up to and including the level of the cover. In this regard, if the insured is sued then it is normal practice for the insurance company exercising its rights of subrogation to decide who will represent the insured in defending the claim, and subject to the terms of the policy, the insurers then ought to meet the legal costs of any such action. In addition the insured must not take steps to prejudice the insurers' position. Difficulties can arise with the insured's rights of subrogation and the insurers' desire, for commercial reasons, to resolve any dispute quickly and/or on less favourable grounds in the short-term.

Environmental insurance

In the context of environmental "incidents", it is the Public Liability and Environmental Impairment Liability (EIL) policies which are most relevant. Nowadays, the former – if environmental difficulties are not excluded from cover – is restricted, in effect, to one-off incidents. In this way, the EIL policy dovetails with the Public Liability policy by covering gradual and fortuitous although still unexpected and unintended release or escape of substances. It protects the insured against his legal liability and for two consequences, namely, legal liability for third party pollution outside the site and, generally, off site cleaning up costs. It will normally cover damage to third party property and person. One of the major differences between EIL policies and Public Liability policies is that the former will tend to apply, as with Professional Indemnity policies, to claims made against the insured during the period of cover, while the latter will cover accidents occurring during the period of cover. This reflects an important distinction under insurance policies between those that are on a claims made basis and those that are on an occurrence basis. It also emphasises the importance of reading each policy wording correctly. Insurance policies are contracts and the words will vary. By way of example, some environmental insurance policies, such as that offered by Reliance National in the United Kingdom cover sudden and gradual pollution and may contain, for an additional premium, a 12-month extended discovery period to report claims after the policy has expired.

Public Liability policies tend to cover legal liability based upon an accident or an incident. It is comparatively straightforward to define the accident which occurred during the period of cover. Most policies are renewed on an annual basis. If an accident under a Public Liability policy has not occurred within the year of the cover then the policy will lapse with no claims being made and a new policy will be required.

However, where it is difficult to identify the occurrence such as with the gradual process of pollution, problems can arise, particularly with older Public Liability policies which had no environmental exclusion. There may be a considerable time span between events that give rise to pollution and the pollution coming to light or being rectified. Many old Public Liability policies did not seek to exclude environmental risk (this exclusion is to be found in most new policy wordings). Consequently, in the United States in particular, the subject of insurance archaeology has become important as insureds look back at their old policies to see whether a claim under Public Liability could be made in respect of a "pollution occurrence". Even if there were cover, however, it would tend to be restricted to third party liability only as most old Public Liability policies would not cover clean-up costs.

One of the interesting aspects of insurance archaeology is whether the US lessons will also have to be learnt within the European framework. For example, US law tells us that there can be as many as four different tests applicable in determining what is or is not a relevant "occurrence" under a policy, namely: the "exposure" theory, the "manifestation" theory, the "injury in fact" theory and the "triple trigger" theory. Here is not the place to consider each of these approaches in detail. But what the US experience does teach us is that even on comparatively straightforward wording there can be uncertainty as insurers dive for cover in the face of a major pay-out.

Many EIL policies are now on a claims-made basis which is similar to Professional Indemnity. They cover liability for claims made against the insured during the period of cover. They avoid the long tail of the incidents-occurring basis which has become unpopular with insurers. However, one must be careful with a claims-made wording basis with the inclusion of a reported policy obligation. This may involve both the incident that gives rise to the claim and the claim itself occurring during the same period of insurance in order for the policy in question to bite. By way of example Reliance National's Consultance Environmental Liability Policy is on a "Claims first made against the insured and reported in writing during the Policy Period". This double standard for notification may be difficult to achieve. Policies may also have extended discovery period wording subject to payment of an additional premium. This enables claims to be reported after the original policy has expired.

Generally speaking, in the United Kingdom EIL policies are few and far between. The experience in the United States has been that they were withdrawn after the huge financial losses of the 1970s. The current trend in the United Kingdom is to produce EIL policies relating to specific industries on a site-by-site basis with an environmental audit being a pre-requisite to any insurance cover. EIL cover is, naturally, more expensive than Public Liability and is normally subject to an annual survey by a professionally nominated environmental consultant, the cost of which is borne by the insured. In 1990 the Chemicals Industries Association Scheme known as CEILIF was launched. In 1991 AG Europe (UK) Limited (AIG) began to offer a wording that was similar to a policy developed for the US market. Later that year another facility was launched by Creamer, brokers to the National Association of Waste Disposal Contractors, with specific inclination towards the waste disposal industry. However in January 1991 the CEILIF facility was withdrawn, one suspects through an absence of sales and it is understood that the Reliance National policies and a similar wording used by Swiss Re have not been very successful.

EIL wording is limited to "environmental impairment" as defined by the policy. Examples include the AIG wording which covers the discharge, dispersal, release or escape of smoke, vapours, soot, fumes, acids, alkalis, toxic chemicals, liquids, gases, waste materials or other irritants, contaminants or pollutants into land, the atmosphere or water, while the Creamer definition covers pollution or contamination of the atmosphere or any water, land or other tangible property or the generation of smell, noises, vibrations, light, electricity, radiation, changes in temperature or any other sensory phenomena but not fire or explosion. Reliance National offers a number of environmental policies. Examples are for Pollution Legal liability (Reliance's environmental impairment insurance for pollution conditions which are location based and result in bodily injury or property damage), for pollution clear-up (again location based) and, on similar principles to both, the Contractors Pollution Legal Liability Policy. The core cover is not dissimilar but one must look carefully at the additional words, particularly at the various exclusions and extensions to the cover that is in place. By way of example, the Creamer cover excludes sudden accidental pollution and is, therefore, prepared on the basis of supplementing the cover of Public Liability policies. Care should, therefore, be taken when taking out EIL and Public Liability policies to ensure that they dovetail correctly leaving no gaps in the cover that the insured requires.

To demonstrate, AIG excludes:

> "Any liability arising from pollution conditions emanating from any underground storage tank and associated piping until satisfactory integrity test results are received, approved and on file with the company, such tests having been carried out in a form approved by the company. Coverage will be only provided for those tanks specifically approved and scheduled on to the policy by endorsement. An underground storage tank is any tank, including associated underground piping connected to the tank that has at least 10% of its volume below ground."

Reliance National uses a very similar exclusion. This merely emphasises the importance of reading the policies carefully. This is as true for the amount of cover as it is for the extent of cover. The AIG cover is up to $40m per insured while the Creamer policy provides up to $5m. Reliance National offers up to $5m. Furthermore some insurance policies will offer legal costs and expenses including expenses incurred in defending criminal proceedings against a company or individual – but not, of course, the amount of any criminal fine that may be imposed for breach of environmental regulations (such as under the Environmental Protection Act 1990 and the Water Resources Act 1991) as this would be contrary to public policy. The same type of cover may also be available in Health & Safety prosecutions. Having said that, one should always bear in mind that it may be possible to claim a reimbursement of a criminal fine by way of an express or implied contractual indemnity from a party whose conduct caused the offence: in the case of *Osmond* v *I Ralf Moss Limited* (1970) a motorist was fined for the strict liability offence of driving while uninsured. He was able to recover the amount of the fine from the insurance broker who led him to believe that he was insured. This case may have major repercussions in the event of third parties who give advice on, for example, environmental matters (*e.g.* an environmental auditor) whose negligent advice ultimately leads to a criminal liability and thereafter to civil claims by third parties who have suffered a loss.

Environmental auditors

The environmental auditor when giving advice, would ordinarily seek protection for the liability consequences of negligent advice from his Professional Indemnity policy. It is important to bear in mind that the policy is based upon a legal liability and that what constitutes a legal liability may within the European Community vary from one Member State to another. The risk of being pursued in a country other than that which may be regarded as the professional's home territory is much greater than formerly, now that the Brussels Convention – implemented through the Civil Jurisdiction and Judgments Act 1982 in the United Kingdom – is in regular use.

In the United Kingdom the law of negligence (more fully discussed in Chapter 8) has been in retreat for some time and seeks to distinguish between:

(i) financial losses which are immediately consequent upon physical damage to people or property (such as loss of profit or disruption occasioned by physical damage); and
(ii) other financial losses which are, in the legal mind, wholly unrelated to physical damage.

These latter losses would include the cost of making good quality defects which are not in any way dangerous to neighbouring people or property, investment losses and diminution in the value of assets purchased in reliance on negligent information or advice. The distinction exists, so it is said, because of the need to limit the extent of liability for financial losses. Physical damage, and its direct financial consequences, tend to have natural limits. By contrast financial losses are conceptual only with no necessary relationship to the physical world. Thus there are no natural limits to the extent of the financial losses that could flow from a causative negligent event so far as tortuous negligent liability to third parties is concerned. Contractual liabilities are, by contrast, naturally limited in any event to the parties to the contract (the privity principle). Therefore, contract law does not need to distinguish between physical and financial losses.

The law of negligence does need to impose artificial limits and to control what has been seen as the unacceptable proliferation of claims. Therefore, the courts draw this important distinction between financial losses which are consequent upon physical damage and claims which are purely in respect of economic loss unrelated to physical damage. It is in connection with this latter category that the courts tend now not to allow recovery. But on what legal basis do they draw the distinction? The ordinary and general duty of care in tortious negligence is to take reasonable care to avoid causing physical harm to the person or property of others. However, the problem for the courts has been to decide upon the circumstances in which this principle might be extended. If foreseeability of the harm was the main criterion for identifying losses which were to be compensated by damages in the tort of negligence, then in many instances it is just as foreseeable that third parties will suffer pure economic loss in the cost of making good quality defects or through money lost in reliance on negligent information as it is that they will suffer physical harm. As foreseeability remains a touchstone, the courts have in the past found it increasingly difficult to resist extension of the duty of care from physical harm to economic losses. How did the problem start? In 1963 in *Hedley Burn* v *Heller & Ptnrs* the House of Lords recognised for the first time that in appropriate circumstances a duty could extend to taking care to avoid causing wholly financial loss by negligent information or advice. It is in this area in particular in relation to environmental auditing that one needs to be wary of the law of negligence and the principle of *Hedley Burn* (as developed in *Caparo Industries* v *Dickman*).

The impact of these cases is discussed in some detail in Chapter 8. Their importance in the context of insurance lies in the fact that, first, as we have mentioned, Professional Indemnity cover exists only for legal liabilities and, secondly, that cover generally only exists for what could be tortious liabilities. To the extent that, say, an auditor is exposed to additional liabilities by reason only of his contractual terms of engagement, there will be usually be no cover. So an awareness of the extent of potential tortious responsibilities is critical to ensuring adequate insurance protection.

Future trends in environmental insurance

It is apparent from lessons from the United States that environmental issues do need to be faced and that companies may, by ignoring the issues or not taking steps to deal with them, incur substantial liabilities. Generally speaking, criminal fines are not covered by insurance and, therefore, will come off the top of company accounts. With the arrival of the Environmental Protection Act 1990 and the Water Resources Act 1991 there has been a trend towards higher fines. Although they are limited to £20,000 in the Magistrates Court, there is no financial limit for the more serious offences which are normally dealt with by the Crown Court. To date fines are running into hundreds of thousands of pounds and have reached the million pound mark, particularly when clean-up costs are added to the fines themselves. It is never easy to predict trends but, if the experience of US judicial practice is a guide, then the level of fines is likely to increase still further and the possibility of directors receiving custodial sentences may become more real (for example, under section 157 of the Environmental Protection Act 1990 and the equivalent section 217 of the Water Resources Act 1991). In the United Kingdom, there are embryonic signs of this in the field of health and safety legislation, as noted in Chapter 6.

Against this background of growing liability and financial risk, there is no doubt that, overall, environmental insurance does not offer the salvation that many, perhaps, had hoped. It clearly has a role to play, but the amount of cover and the terms under which it is given will not be able to meet the risks on environmental issues that companies and individuals are likely to face over the next few years. There is clearly much scope for an improvement of the existing insurance policies which deal with both sudden and accidental incidents on the one hand and gradual incidents on the other. It may be that the process of taking out insurance by way of a proposal form will have to be changed and that companies will have to receive some sort of certificate from an independent expert stating that they are complying with present environmental regulations so that it is the expert, rather than the prudent underwriter, who decides whether or not all material facts have been disclosed for the purpose of taking out cover.

Overall it does seem as if EIL, and the approach being adopted for certain industries, and environmental audits may provide some of the answers – as much by comforting those involved as anything. However, there may again be a rise in the use of forms of risk funding (see, for example, the present proposals of NAWDC towards the new UK waste regulations, P&I shipping clubs and mutual insurance arrangements) rather than more typical risk transfer through insurance alone. Rather than the insured and the insurer doing battle over whether or not there is cover, it may be that members of certain industries and their trade associations (and possibly government) should provide an appropriate fund which in certain circumstances could be drawn down in the event of an environmental claim. A common and consistent approach to the concept of identifying the risk (probably in part by way of an audit), and then managing and funding it, appears on the face of it to be a more sensible and complete way forward than present insurance arrangements and the law alone can provide.

Confidentiality - risks and suggestions

Whatever policy in relation to environmental auditing one considers, there is always the obligation to disclose to the underwriter all material facts, *i.e.* all facts that will affect the underwriter's decision as to whether to accept the risk and, if so, on what terms. This duty of disclosure may result in information being produced which could be obtained by third parties, including, at its limits, the public at large. In addition, there is a growing trend towards more accountability and more access to information (*cf.* the Freedom of Access to Information Directive that will be in force by the end of 1992 with the implementation of the Environmental Information Regulations by the UK government (SI No 3240) (see also Chap 4)). An issue that arises is whether this information can remain confidential. The environment is now subject to numerous regulations, a breach of which may initially lead to a criminal prosecution and subsequent fine (not covered by insurance). Thereafter, a civil action may follow with claims by third parties who may have been damaged or suffered a loss (possibly covered by insurance). Is it appropriate for the insured and the insurer to consider these risks under the umbrella of confidentiality or is the seriousness of damage to the environment such that the public at large have a right to know? Can confidentiality arrangements be maintained in the face of legal proceedings, particularly in the United Kingdom with its notorious system of systematic discovery of documents?

Looking at the issue of confidentiality more widely, it is plainly the case that, as a matter of good management practice, companies will wish to establish procedures for dealing with what are broadly described as environmental matters and these will more often than not be set down in writing in one form or another. They may range from notes on environmental performance to comprehensive statements of policy, from internal guidelines of an entirely general nature to identified steps that are recognised as needing to be taken to protect adjoining property and third parties from pollution. In addition, there may be reviews in the form of board meetings of the company's progress to date in improving its environmental performance including reviews based upon information generated in the course of operating the EC Eco-audit scheme.

The objectives reflected in the preparation of these documents will vary from those that are purely practical to a degree of marketing and public relations activities both internal and external. Over the last few years the public's knowledge of environmental issues has grown considerably. Consumers now purchase "green products" and the use of market forces is becoming more and more used by companies and countries to gain a market edge. It would be surprising if, by way of example, house builders did not market houses not built on contaminated land as being a better purchase than those that are built on sites with a former contaminative use. This is despite whatever effective clean-up may have taken place, because consumers' present perception of "contaminated land" is generally not conducive to a house builder in achieving a subsequent sale. It is against this background that care must be taken over the production and method of production of documents, such as those produced during an Eco-audit or any other form of environmental audit. Internal environmental audits commonly reveal dangers that have been identified within an operating system long before they were in fact the cause of any incident or accident. These documents are inevitably likely to be discoverable.

How can the risks of these various documents be controlled? The first point is that statutory obligations regarding production must be complied with. However, there are a number of documents which one may wish to word cautiously and for which one may wish to restrict the likely circulation. There is, of course, a balance to be struck between recognising the statutory, management and marketing purposes of documents (namely the desire to show a sufficiently comprehensive approach to environmental matters) while avoiding going beyond what is necessary and steering clear of over-enthusiastic assertions which may subsequently be turned against the company.

One possibility may lie in the legal professional privilege exception to the general rule on disclosure of documents in common law systems which is that, generally speaking, most documents are discoverable if they are relevant to any issues in dispute and are in the custody, power or control of the plaintiff or defendant. To establish legal professional privilege, it will generally have to be the case that the sole or dominant purpose of the document is to seek legal advice. In many cases litigation will not be pending so that what is generally regarded as the second limb of the exception is not very helpful, namely that for documents prepared in contemplation of proceedings. The third exception relates to government public policy privilege.

In order to show that the dominant purpose of the report is to seek legal advice, care must be taken in structuring the document and structuring the organisation so that lawyers, external or internal, are involved in the investigation and reporting process. Precisely how that can best be done will depend upon each company's particular circumstance. It is important to remember that whether a document is privileged or not depends upon the substance of its content and the purpose for which it is prepared. It should not be assumed that some self-evidently formal step such as merely copying the document to a lawyer will suffice to render it privileged. Equally one should not assume that a report on an environmental accident for insurers only will attract privilege as the overriding test is whether the dominant purpose of the document is to seek legal advice.

It is important to remember that as companies through insurance, public pressure, shareholder demands and employee demands are more and more committed to undertaking environmental auditing, this will produce many documents, the majority of which may be disclosed to third parties. The effect of a document prepared in one context being used in an entirely different manner from that which was intended or is necessary for it to be properly understood is well known. Equally, companies should not be lulled into a false sense of security by the present benevolent approach of some of the regulatory authorities in allowing companies to withhold information by reason of commercial (not legal) confidentiality. One anticipates that many legislators (particularly from Europe) did not intend this and will, with growing public support, seek to restrict what many environmentalists perceive to be an "information loop-hole".

The management systems that companies choose to operate and the way in which documents are put together are of great significance. It may be that companies will decide not only to introduce a lawyer into their management structure but also to prepare two documents rather than one, the second being dominated by its purpose to obtain legal advice and which – understandably, given what may be the perceived likelihood of legal proceedings – contains details of the more sensitive areas of a company's environmental performance. However, it is just these sensitive areas that in any event must be disclosed to insurance companies for the purpose of obtaining cover. Striking a balance is not something that is likely to be easy. The difficulty for any company which seeks to withhold information, is that those learning of it thereafter will almost certainly consider that the company has something to hide.

CONTENTS OF CHAPTER 10

Eco-auditing and Information Technology

Chapter 10
Eco-auditing and Information Technology

The purpose of this chapter is to consider a number of ways in which information technology (IT) can support and enhance the process of Eco-auditing. One immediate attraction of using IT in this context is that Eco-auditing is itself in its formative stages, so computers might be introduced at the outset as a way of undertaking an Eco-audit within many organisations. This would contrast with the more conventional deployment of IT, which usually involves the imposition of computer technology on a pre-existing manual task or activity often ill-suited to automation.

More fundamentally, however, the process of Eco-auditing is essentially an information processing task. The conduct of an Eco-audit presupposes access to a wide body of information, relating both to the requirements of the audit, as well as to the individual features of the organisation being reviewed; and the audit itself demands the systematic and objective application of a number of information criteria. Given that Eco-auditing is both information-dependent and information-driven, there could be clear applications of IT both to automate the process and perhaps also to give rise to more imaginative and creative ways of conducting audits.

This chapter is intended to provide an overview of a number of enabling technologies and, with the examples presented, gives an indication of the potential benefits and the management challenges that face organisations considering the development of computer-assisted Eco-auditing systems. The chapter deals with the following topics:

- environmental information systems;
- intelligent check-lists;
- audit automation;
- diagnostic and planning tools;
- users and developers;
- user interface; and
- project management issues.

Environmental information systems

To conduct an Eco-audit efficiently, auditors need speedy and easy access to large amounts of information. Similarly, individuals within organisations responsible for ensuring compliance with environmental regulations also need to be able to retrieve relevant information. The information being considered here is not the data specific to particular organisations (although that could also valuably be stored and disseminated through IT), but the quantities of regulation and background information that underpin this area. Examples of such information include European Directives, national legislation, internationally recognised standards and all manner of rules and regulations governing particular trades and industries. Additionally, auditors and compliance officers are faced with daunting and growing bodies of secondary information that relates to and clarifies the application of the regulations and standards. Such secondary information includes consultation papers, newsletters, and learned articles.

It is indeed a formidable task for anyone undertaking an Eco-audit or charged with ensuring compliance with environmental regulations to master these ever-expanding bodies of information, and it is precisely in coping with this challenge that IT can help. In particular, there are two enabling technologies which, in isolation or combination, can greatly reduce the burdens of information management. These are full text retrieval systems and hypertext. (There are, of course, other relevant technologies but these two have been identified as particularly well suited to the mix of regulatory and background information that needs to be handled.) These technologies would not themselves automate the Eco-audit process (such systems are considered later in this chapter) but would assist the task of conducting an audit.

Turning first to full text retrieval systems, the idea here, as the name suggests, is to store documents in such a way that they can be retrieved by instructing the computer to search through the text for the occurrence of words. These words are often referred to as "keywords" and can be searched for in isolation or in combination. For example, the user of a system that held large quantities of regulatory and background information on environmental matters, could require it to search for all occurrences of the word "packaging", and the system would then identify all those documents in which that term appeared. Alternatively, or additionally, the user may wish to search for those documents in which the word packaging occurs within a specified number of words of another term such as "disposable". Through a number of techniques, the searches can be refined and achieve considerable sophistication, allowing for the conjunction, disjunction and other combinations of words.

Full text retrieval systems have a strong history in the legal profession. Perhaps the best known system is LEXIS, one of the largest databases in the world. This system spans many jurisdictions and holds huge quantities of primary and secondary legislation, case reports and secondary writings as well. The LEXIS system, perhaps more than any other, has confirmed the potential of managing large amounts of regulatory information using full text retrieval techniques. There are a number of European databases which operate under similar principles, for example the CELEX system.

Whatever particular techniques or technologies are used, however, the underlying motivation is the same: to make textual information available in searchable format so that users are able to identify documents on the basis of words which occur within them. While this has been proven to be a powerful technique, it must be conceded that this method of accessing documents by keyword search requests is not without its problems. In particular, sloppy searching or naive users can result in searches that are formulated too widely and retrieve an excess of documents, many of which are irrelevant, while too restrictive a selection of keywords can result in a failure to identify all relevant documents. In response to these shortcomings, of both system and users, a number of techniques have been developed. Some of these help the user in the task of actually formulating search requests, while others engage in what is known as "conceptual searching", which involves the system searching not just for the occurrence of words stipulated by the user, but also for related terms which the system has been programmed to identify as conceptually similar. These techniques are gradually being introduced to the packages that are popularly available, even to run on personal computers, and full text retrieval systems can be expected to improve in performance over the coming years.

The second technology that can help in the straightforward management of regulatory and background information is known as "hypertext". This is far less widely used than full text retrieval technology and it is considered by its main proponents to be as powerful a tool for lawyers and businessmen as the spreadsheet has been for accountants. Hypertext technology provides an ideal mechanism for linking documents to one another, cross-referencing between documents providing context-sensitive help, and generally offering a navigational tool through which users can browse around legal and related information. Full implementation of hypertext involves not simply the ability to display relevant pieces of text on screen, but actually to navigate freely around a substantial and highly complex information space. As a means of structuring regulatory material, of providing commentary on such material and of pulling related material into one easily accessible information source, hypertext stands out as the most likely candidate for exploitation in this decade.

Hypertext can be likened to an "electronic book" and often appeals to non-technical users in that it allows for the intuitive and simple navigation around documents. Whilst the effort involved in setting up hypertext systems can be significant, it would be ideally suited to the large, complex databases of relatively static information which the field of Eco-audit is set to spawn. The availability of information systems holding environmental information would be of great benefit to auditors and compliance officers alike, allowing and encouraging them to roam around materials, both formal and informal, in search of relevant information.

An illustration of the application of hypertext technology is the FIDIC system developed by Masons in conjunction with the International Chamber of Commerce in Paris. This system holds the full text of Parts I and II of FIDIC, the international standard form construction and engineering contract. The system presents the text of FIDIC on the screen in such a way that it requires no technical expertise to browse through. It offers users a number of facilities. For example, if a term is contractually defined within the agreement, then the definition of the term can be called up onto the screen simply by placing a cursor over the relevant word in the text and clicking on the button of the mouse. Similarly, whenever there is a cross-reference to another clause, then the user can jump to this part of the contract, again just by pointing and clicking. Other information is also held in readily accessible form, so that a user wanting explanation of the clauses of the contract can, again, at the touch of a button, call up context-sensitive expert commentary prepared by a specialist that will clarify each clause and sub-clause, as required by the user.

There is also the possibility of combining full text retrieval systems with hypertext, so that a user might search in the first instance for documents on the basis of a keyword occurrence, and having identified some relevant documents be able thereafter to jump easily into other relevant documents that have been linked to the retrieved materials in anticipation of the needs of users.

Intelligent check-lists

Many Eco-auditors or compliance officers may wish to go beyond having published information available to them on an electronic basis and receive more systematic and structured guidance on the conduct of audits or the review of their organisations. There is a way of achieving this further guidance from technology, yet falling short of the fully developed audit tool that is discussed in the following section. This more modest approach takes advantage of a technique that can be referred to as the "intelligent check-list".

Many complex tasks facing auditors and managers require extensive expertise and knowledge which have, in the past, often been encapsulated as notes and procedures in paper check-lists. Intelligent check-lists are computerised implementations of these paper check-lists but are easier to use, store and maintain. They can take their users through complex and extended procedures, ensuring that all matters are attended to – and dealt with, if relevant, within a prescribed time period. For example, an intelligent check-list could assist a user in auditing and reviewing compliance with some specific piece of environmental legislation.

Compliance reviews must be undertaken with relentless attention to detail and extensive reference to regulations. While paper check-lists have been found helpful in the past for such tasks, it has been a common experience to mislay the precise paper check-list required at any given time.

The attraction of holding check-lists in electronic form is twofold. First, they can be held securely in one place and be constantly available for easy access. Secondly, the systems can hide complexity from users, in that, at any one time, the user is presented with one screen of information containing elements of a check-list, hidden beneath each of which will be further information and perhaps other check-lists. Rather than being faced with long and daunting paper check-lists, much of which is often not relevant for the particular tasks of individual readers, the intelligent check-list approach exposes users only to those issues which they select as relevant and thereby renders the review process more manageable.

The technique of intelligent check-listing is most easily delivered using hypertext, the enabling technology discussed in the previous section. In operation, therefore, such a system will present a series of items on the screen and further information on each can be obtained by pointing and clicking on the topic of most interest to the user. The check-lists can also have internal cross-referencing and the ability to roam from item to item at the user's will.

Intelligent check-lists can also be integrated with environmental information systems of the sort described in the previous section, so that, for example, a user browsing through a particular piece of environmental legislation may be informed that an additional facility is available (*i.e.* a check-list) which will offer structured guidance to the user whose task involves considering various aspects of compliance with the regulations in a controlled and rigorous way.

Audit automation

Perhaps the ideal tool for Eco-auditors or compliance officers would be a system that asked a series of questions regarding a company under consideration and, on the basis of the user's responses, would undertake a full Eco-audit. This echoes one of the challenges that has faced major financial audit practices over the last 20 years or so: the full automation of the audit process. All the world's leading accounting firms are investing heavily in moving towards this goal but progress has not been as substantial as was anticipated in the early 1980s.

One of the major hurdles in the development of automated audit systems (both for financial and Eco-audits) is the current state of the most relevant technology: that of expert systems. To understand the potential of audit automation systems, it is helpful to appreciate the background to expert systems technology.

Research, development and implementation of expert systems is under way in many commercial organisations today. The popular and academic press abound with commentaries on artificial intelligence (AI) and expert systems. Yet for many people, the basic ideas and terminology of these fields remain confused and esoteric.

Broadly speaking, AI programmers seek to design, develop, and implement computer systems that can perform tasks and solve problems of a sort for which human intelligence is usually thought to be required. For instance, AI programs are being written that understand the spoken word, translate limited subsets of natural language, and recognise objects of the physical world.

Expert systems are one sort of AI program. These are systems designed to reason, solve problems, draw inferences and offer advice in an apparently intelligent fashion. In medicine, for instance, expert systems assist in the process of diagnosis, while in computer manufacturing and sales they can offer advice on the configuration of systems.

Expert systems are built with the assistance of human experts – in medicine, law or whatever area of application – in such a way that they are capable of drawing conclusions and solving highly specialised problems (normally within limited fields of discourse). They are designed to reason at the level of, and sometimes at an even higher standard than, human experts in their given domains. Much as human beings do, expert systems offer explanations for thin lines of reasoning and are also capable of reasoning with incomplete or incorrect data by drawing tentative conclusions. So, within limited domains, if someone is confronted with a problem requiring expertise for its resolution, rather than always referring it to a human expert he may instead consult his expert system which will guide him through the problem solving process. In this way, the valuable and scarce resource of human expertise can be more efficiently allocated.

The conduct of an Eco-audit is precisely the type of activity that requires expertise which is often in scarce supply. The challenge for the world of expert systems is to provide technologies that will allow the expertise underlying Eco-auditing to be embodied in a system accessible to less expert users undertaking an audit.

One of the problems that has faced those seeking to automate the financial audit process has been the range of knowledge and expertise required to conduct the full audit (in anticipation of the vast spread of financial scenarios that can arise). There is here a significant contrast with the process of Eco-audit, which seems to depend on a more self-contained and manageable body of regulation and expertise. On the face of it, it will be easier, to use the jargon of expert systems, to "engineer the knowledge" of specialist Eco-auditors than that of financial auditors.

In this respect, there are similarities between the field of Eco-audit and that of the domain chosen for one of the United Kingdom's best known expert systems, VATIA, developed originally by Ernst & Whinney in 1988 to automate the process of auditing a client's VAT affairs. Ernst & Whinney's response to the new VAT regulations, which were at once extremely complex and many in number, was not to recruit large teams of VAT specialists, nor to require the use of still more complicated paper check-lists by general auditors, but rather to develop an expert system that would allow the firm's general auditors to tap into the knowledge and expertise available within its VAT specialist group without having to refer every point of difficulty or doubt to them. The system was designed to be used in the presence of clients, as "an intelligent assistant", while appraising the adequacy and compliance of clients' VAT systems and procedures. The combined knowledge of the users and the system together produced an effective and thorough scrutiny of clients' VAT positions with minimal direct involvement of the firm's VAT specialists. The system ran on the firm's standard audit microcomputer, which at that time was a portable IBM compatible machine.

Similarly, it is entirely conceivable that a system could be developed to support the process of Eco-auditing. It would guide its users through the complex webs of environmental law and regulation, exposing them only to details relevant to the audit in hand. The direction of the search through the knowledge base, the system's store of expertise, would be conditioned largely by the nature of the business under scrutiny, a matter that would be addressed at the outset of any consultation with the system. The user would interact with the system by answering "yes", "no", or "don't know", by selecting from menus or by entering data.

The system could be designed so that a record of any consultation would be retained, both on paper and on disk. Answers supplied during any consultation could be stored on disk and so be available to a reporting program, or as input for a subsequent run of the system. This would enable later use of the system (perhaps the following year) to be carried out against answers provided during earlier consultations.

It is likely that consultations with such a system would take up to an hour, at the end of which the system could provide a printed summary of all areas of concern that had arisen. These could include points for senior management to note, points where immediate action is required and points for auditors to note. Throughout consultation with such a system, context-sensitive help screens could be available to assist the user in answering the questions. Again, such a system could be integrated with the environmental information systems mentioned earlier. In operation, this could mean that the user of the system could suspend the current consultation and perhaps navigate through a series of informative screens linked through hypertext.

The development of an automated Eco-audit system of the sort outlined above is by no means a trivial task. In terms of human resources, the power of the system is ultimately dependent on the effort expended by the expert whose knowledge is held in the package. The knowledge engineering process is an arduous and time-consuming job and, without the commitment and enthusiasm of a willing expert, it is unlikely that a system would be delivered within a sensible time-scale. As well as the expert's input, there is a need for the kind of specialist who develops such systems. Known as the "knowledge engineer", his role involves the ability both to help elicit the expertise from the specialist and then to articulate and represent that knowledge in a form that can be held within a computer system. The domain expert and knowledge engineer would need to work hand in hand, probably for many months, to develop such a system.

Diagnostic and planning tools

There are further categories of system that would be of help to those involved with Eco-auditing. These would go beyond the facilities for browsing offered by environmental systems and the structured guidance offered by the intelligent check-lists; and even beyond the audit systems that would undertake the Eco-audit itself. The more advanced systems would be devoted to the diagnosis of specific, highly specialised legal issues and to the provision of assistance to users involved with the detailed planning of a company's affairs in order to comply with regulations and to maximise commercial opportunity. The systems being described are further categories of diagnostic and planning systems, respectively. This section discusses each category in turn.

Diagnostic systems offer specific solutions to problems presented to them. Such a system itself will help elicit the facts or circumstances of some actual case or scenario, and will analyse the details and draw a conclusion, usually after some kind of interactive consultation. These systems are perhaps best exemplified by the medical diagnostic systems which make diagnoses on the basis of symptoms presented to them. The best known example of diagnostic systems in the legislative area are taxation systems that can pinpoint exactly the extent to which – and why – a person is liable to pay tax, doing so on the basis of a mass of details provided by the operator.

Although it is technically possible to develop such systems, very few have yet been made commercially available. The world's first published system was the Latent Damage System, which offers a clear indication of the potential of this technology. The system itself offers advice and draws conclusions in relation to the Latent Damage Act 1986 and related statutes and judicial precedents. It is for use by lawyers and legally informed people. The system is an expert system and it advises on the law on latent damage, dealing fully with the Latent Damage Act 1986 and the complex context of tort, contract and product liability law in which that piece of legislation operates. This branch of law has wide-ranging and recurrent implications, not only for all professional advisers (solicitors, accountants, consultants, and so forth), but also for the manufacturing and construction supply industries. The purpose of the package is to assist in identifying when a breach of duty occurred, on what date a cause of action accrued, and when the wronged party will be taken to have had knowledge of the loss suffered. The implications of fraud, deliberate concealment, contractual liability, personal injuries, product liabilities, successive ownership and many other related matters are also taken into account. The key piece of advice offered by the system is the date after which a claimant can no longer commence proceedings. The discovery of this date is the result of a sophisticated reasoning process.

In highly specialised and technical areas of environmental law it is likely that the technology used for the Latent Damage System could be usefully employed to clarify with speed the interpretation of particularly complex regulatory areas. Once more, there is great potential for integration: as before, it would be possible to have an audit automation system, enhanced by environmental information systems as well as by diagnostic system modules. There would be still further scope for integration by introducing the second category of advanced system mentioned earlier, namely the planning system.

Planning systems reason in the reverse direction from diagnostic systems, for they are instructed as to a desired solution or outcome, and their purpose is to identify scenarios, involving both factual and legal premises, that justify the preferred conclusion. Most of the pioneering work in this area, again, was conducted in the field of tax law, where planning systems can recommend how best a taxpayer should arrange his affairs so as to minimise his exposure to liability. So the knowledge held within planning systems can be very similar to that held within diagnostic systems; what is quite different is the way that that knowledge is applied.

A leading example of a planning system is THUMPER, an expert system that advises on Stamp Duty. The user of THUMPER tells the system what commercial transaction is under consideration, such as the acquisition of a business. In turn, the user is invited to indicate only those matters that bear on the transaction under consideration and which may affect liability for stamp duty. The system analyses the facts presented to it and generates a report on the stamp duty payable, referring also to any relevant exemptions, and as trends emerge, the system is able to provide planning advice.

Similar technology could be used in conjunction with findings of the Eco-audit process. Such a system could advise on ways in which a company could organise its affairs in order both to comply with the requirements of Eco-auditing and to realise other business benefits. Such a system, combining audit and planning features, could be most useful in the context of considering an acquisition or merger. One key factor emerging in the appraisal of target organisations is current and foreseeable compliance with environmental regulations.

National Westminster Bank has recently released its PHAROS system, which combines the diagnostic and planning techniques outlined above. The system is intended to assist its users in assessing how their businesses may be affected by the creation of the European Single Market; and in deciding on what strategic and management decisions should be made. The system is designed for use by UK businesses, especially for companies with an annual turnover of between £0.5 million and £150 million.

It is interesting to note that the first update to the PHAROS system includes a module that advises on environmental matters. This is designed to help users consider how their businesses are affected by recent environmental legislation (for example, the Environmental Protection Act 1990) as well as relevant Directives (for example, the Services Directive and the Toy Safety Directive).

It is useful to put the PHAROS system in the context of the various technologies discussed in this chapter. Its main purpose is to analyse and diagnose the company's position in relation to large bodies of relevant regulation. Additionally, there are some features which assist in the planning process. The system does not offer the facilities offered by the environmental information systems as described here, nor indeed of the audit automation tools discussed above.

Users and developers

The systems introduced and described in this chapter could be developed and used by a wide variety of individuals and organisations.

Regarding potential users of the system, all the applications envisaged would require their users to have some knowledge of Eco-auditing and environmental regulation generally. In relation to the environmental information systems, the user of a full text retrieval system would need to have some sense of the terminology relating to the subject matter into which he was enquiring, while in the hypertext system, to avoid getting lost in what is often metaphorically termed "hyperspace", the user must have some broad sense of the environmental domain. As regards intelligent check-lists, while these would no doubt provide useful prompts and guidance for users, they would not be intended for someone who has no knowledge whatsoever of the process being described. An even greater level of knowledge would be required of the user of an automated audit system, in that any question posed will inevitably assume the user has some broad understanding of the environmental context, and indeed the business context, in which questions are asked. Finally, even with the diagnostic and planning tools, no matter how sophisticated these aids are, they will none the less require that their users are able to understand questions that may often themselves be couched in fairly specialised language.

In terms of specific categories of user, clearly Eco-auditors themselves are very strong candidates. Such auditors may work for external audit organisations, they may hold themselves out as "environmental" auditors and soon they may be "accredited environmental verifiers", the new profession of practitioners which could be set up under the EC Eco-audit Regulation. Equally, these auditors may operate purely internally, analogous to internal financial auditors, both to ensure that the company meets corporate environmental objectives as well as to plan for and pre-empt any problems that might arise when external Eco-auditors are called in.

There are a wide variety of management consultants who may also be interested in computer-assisted Eco-auditing. Environmental specialists, strategic consultants, merger and acquisition advisers and risk assessment professionals would all benefit from the regulatory advice and information sources that technology could offer, in the ways described in this chapter. These advisers would go beyond the Eco-audit specialists, in that the latter would tend to focus solely on environmental matters, where the former would place the findings of the system in the wider business context in which they had been instructed. One of the challenges for these more general consultants, therefore, will be to integrate any Eco-audit systems with the other technologies that they use.

Another important category of user would be top management. Matters relating to the environment and compliance with relevant regulations are, for many European companies, climbing up the agenda at board level within companies. Top managers who themselves may seek in-depth personal understanding of the issues, perhaps for promotional purposes or perhaps as part of their strategic review process, would benefit from access to a variety of the systems outlined here. One further use in this context is for environmental specialists to work hand-in-hand with top managers, using a consultation with a system as the focal point of their discussions. Such reviews can also be conducted valuably in group sessions by projecting the image from a personal computer onto a wall screen, again providing a focal point for environmental debate.

Just as there are many potential users of computer-assisted Eco-auditing, there is also a large number of likely organisations who could fund and support the development of these systems. It is conceivable that the European Commission itself, or the European Environmental Agency, once established, would wish to back a project devoted to the development of these applications. Within individual countries, there may also be agencies willing to participate in relevant development projects – this would be a salutary project, for example, for Ireland's Environment Protection Agency or perhaps the Ministry of the Environment in France. Equally, industry groups or trade associations within countries and across Europe could well have a collective interest in collaborating in projects devoted to producing systems for computer-assisted Eco-audit.

Additionally, consultancy firms and specialist auditors will no doubt already have invested in the development of some of these systems. The extent to which these applications will be made available beyond these organisations and for the benefit of non-clients of these firms, or indeed their likelihood of hearing about particular systems, remains uncertain.

Finally, the great upsurge of interest by the legal profession in information technology suggests that it is likely that lawyers will become major users of computer-assisted auditing systems and indeed may champion their development. Traditionally, auditors and management consultants have been far greater users of IT than lawyers, and have, therefore, been able to provide a range of services beyond the capability of legal practices. In the United Kingdom at least, this trend is not set to continue. The large accounting/consultancy firms are now consolidating their investment in information technology, while many of the law firms have far more ambitious plans. This in many ways reflects the different types of work that accountants and lawyers undertake – early computing was well-suited to the automation of much of the numerical work of accountants, while more recent technical developments have allowed more effective processing of textual information which is more closely related to the task of lawyers.

The user interface

Perhaps the key component of any computer system for direct use by professionals is the user interface. This is the part of the system that controls the way in which the user can interact with the technology: what the system looks like on screen, and what a user needs to do (for example, press a key or click on a mouse) to get the system to do what is wanted. For most professionals, the user interface must be simple to understand and intuitively obvious in operation. If it is cumbersome or complex then – no matter how sophisticated the rest of the set-up might be – the system will invariably fail to meet business objectives.

Traditionally, computer systems have been horrendous from the user's perspective. This has not affected or worried most professionals too much because computers have rarely been used by professionals themselves but have been available for secretaries and support staff. However, as machines have appeared on professionals' desks, user-friendliness has jumped rapidly up the IT agenda in service organisations.

There has been a simultaneous shift in industry generally towards recognition that computer systems are in fact mainly for use by average mortals, and not just for computer specialists. In the words of an American commentator, David Liddle, it has now been acknowledged that:

> "It's time to get computing out of people's way. Those who are proud of their ability to withstand pain can continue to do computing. The rest of them need to be given a user interface that lets them go ahead and do their job."

In the past, user interfaces were character-based, while, more recently, there has been a radical shift towards a graphical user interface (GUI). The fundamental difference between GUI-based systems and character-based systems is that the former depend on recognition while the latter require recall: users of GUI-based systems simply recognise symbols that intuitively represent commands or instructions they wish to implement, whereas in character-based systems users are required to recall (often counter-intuitive) words or commands to get the system to do what they want. For example, to start a word processing package using a GUI-based system, the user may need only to "point and click" at a little picture that denotes word processing (perhaps a hand with a pen), whereas the user of a character-based system may need to type some such command as "c:/wp/wp", which is far from memorable and intuitive.

More generally, however, given professional working practices of the past, combined with their general lack of training in IT (and often a related fear or scepticism), it is recognised as crucial that the user interface of any system in no way discourages usage of the system. It would seem, therefore, that the systems discussed in this paper are more likely to be used, and used productively, if they are GUI-based.

So what are the benefits in using GUI-based Eco-audit systems as compared with character-based systems? Generally, GUI-based systems are easier to use; quicker to learn; facilitate more effective combination of applications; and can more faithfully represent (visually) original versions of documents. Each of these advantages are explained more fully below.

In terms of ease of use, the GUI tends to incorporate features described by the rather remarkable acronym WIMP, which refers to the screen format being dominated by movable (often overlapping) windows, the use of icons to represent commands and instructions to the system, the use of a pointing device known as a mouse and the availability of pull-down menus displaying options to users. All of these features contribute to more intuitive operation of computer systems than was possible in the past. Because GUI-based systems do not require their users to memorise cumbersome, convoluted and counter-intuitive commands, but instead simply to recognise symbols and progress through natural steps, professional advisers generally find these systems far easier to use.

With regard to the speed of learning to use systems, many studies by computer scientists and cognitive psychologists have suggested that the use of the GUI also reduces the learning time of users faced with new applications, and, in turn, leads to use of more application packages per machine than when there is a character-based interface. The IT industry itself has recognised the popularity of GUI: UNIX is moving to X/Open or Motif and X-Windows; the Personal Computer (PC) is moving to Windows and/or OS/2; the Macintosh already has an advanced graphical user interface; while the impressive NeXT machine, hailed by many commentators as a pointer to the future, has a particularly sophisticated GUI. For professional advisers, once they have learned one GUI-based system, experience suggests that learning other graphical systems will be a quicker and less painful process and will result in users making use of many more applications than would be the case where the systems are character-based.

A further advantage in using GUI-based systems is that data can easily be transferred from one application to another. For example, a lawyer using an environmental information system within a GUI environment, when analysing a document, can easily "cut and paste" from the database in which he might be searching into a compatible word processing package. Indeed, the screen can often be split so that the database appears on one side and the word processing package on the other. An even more advanced feature common to some graphical packages is the ability to allow data to be dynamically exchanged and updated. For example, a table from a piece of legislation may be held in a graphical database and linked dynamically to the same table in a word processing document, such that the updating of one will automatically revise the other.

The final user interface issue addressed here concerns the way in which the text itself appears on the monitor. One drawback with conventional systems is that the text on the screen often bears little or no resemblance – in terms of page layout, typeface and colour – to the original text in the typewritten source materials. Where there has been a conscious attempt in the development of systems to render the text on the screen identical in layout, font and colour to the original documents, there is often a forceful and positive reaction by users to this and they seem immediately more comfortable with a system whose contents visually resemble the materials with which they would deal manually.

Again, the use of GUI-based systems facilitates such a rendition of the text, because there is an additional benefit of the GUI in terms of what is sometimes known as WYSIWYG (What You See Is What You Get). The GUI's accurate font rendering and use of colour allows documents to be displayed almost exactly as they appear in hard copy. This, in turn, seems to give many professional advisers exactly what they need in relation to computers – a feeling of comfort, familiarity and confidence that they are not in an entirely alien world.

Project management issues

The development of any of the categories of system discussed in this Chapter is an ambitious project and must be managed rigorously if it is to achieve its objective. In particular, project managers should be aware of some specific organisational, technical and commercial issues.

From an organisational point of view, whether the system is being developed by a government department, industry group or independent consulting firm, as elsewhere in the world of information technology, it must be appreciated that if systems are to be successfully implemented within organisations, there must be commitment from top management. The leading figures in an organisation should personally take a firm hold of information technology and stay continually in touch with all major operational systems, upgrades and new developments. The development of the types of system discussed here is precisely the kind of application of which top managers should be aware. The function, price, strategic significance and organisational suitability of any system can and should be appreciated at the highest level within an organisation. In terms of costs and benefits, there is a great deal at stake and there is no justification for passing exclusive responsibility down to the basement or along to the back room. Nevertheless, top management will need technical and administrative support.

A problem here is the lack of appropriately qualified IT managers to take on and run the support function. This human resource problem is symptomatic of much of European industry's use of IT. In an important study recently conducted by the British Computer Society, it was forcefully argued that exploitation of IT will be modest until "hybrid managers" are in place in IT departments within organisations. These hybrids will not only be technical experts (like most IT managers today), but will also (quite unlike IT managers today) have comprehensive understanding of the business which they are supporting, as well as the necessary organisational, personal and management skills needed to maintain the interest and respect of people within their company. Hybrid managers are needed to ensure the successful introduction of systems such as Eco-audit systems, but the management challenges are beyond the capabilities of most computer technicians.

A challenge for both top managers and hybrid managers is what is known as "expectation management". While the systems described in this chapter have considerable attraction and could bring substantial business benefits, it is crucial, prior to their introduction, that the expectations of potential users are managed realistically. Computer systems are never the universal panacea that many marketing specialists would have us believe. Accordingly, the challenge here for IT management is to manage everyone's expectations cautiously, to remain enthusiastic and forceful about the systems but frank and firm about what can reasonably be achieved, given the time and finance available.

Another crucial organisational factor is user involvement. The appearance on the desk top of many professionals of a "2 megabyte internal memory, 80 megabyte hard disk, 386 machine with a fully integrated suite of computer assisted Eco-audit tools" is not necessarily a development that all by nature are likely to get excited about. To reap the rewards of technological innovations, users must be educated to understand and accept the potential benefits. More than this, users *must* want the systems. This means they must constantly be consulted, treated with respect and listened to. Their requirements must be implemented (if it is commercially, technically and organisationally feasible to do so). A marvellously designed system that no-one uses is as much a failure as one that does not work at all. Early involvement of the user and, if possible, clients who will benefit from systems, is strongly recommended.

A final organisational challenge is that of training. There is a tendency across the world of computing to assume the system has delivered all its foreseen benefits once it is implemented in its target office. There is a strong history of failing to invest sufficiently in training users to feel comfortable with the technology given to them. It is essential, therefore, that project managers allow in their budget for the time and finance necessary for the introduction of these systems.

From a technical point of view, given the complexity of the systems, it is important that they are developed in accordance with well-proven methods and standards of system development. In the early days of programming, the development of systems was haphazard and *ad hoc*, with facilities and features being added and removed rather arbitrarily as projects progressed. More recently, however, there has been a move to the introduction and usage of fairly formal system development methodologies, which offer detailed guidance on how to structure and manage complex development projects. In the area of expert systems and hypertext, it is particularly important that systems are fully documented and that projects are subdivided into appropriate phases.

One further useful technique is the development of small prototypes which can give users, clients, managers and experts some sense of what the system will "look and feel" like in operation. On the basis of perhaps just a few days of preparation of this prototype, many more sensible judgments about the viability of the project can be made.

There is often a temptation amongst technical staff, when confronted with the development of systems of the sort described here, to embark on work that involves writing programs from scratch. This temptation must be resisted, as there is no justification for re-inventing the wheel. There are readily available off-the-shelf packages to help organisations build full text retrieval systems, hypertext applications, intelligent check-lists, audit applications, and diagnostic and planning systems. The challenge of system development in this area should, essentially, be to insert the relevant knowledge and information into these shells, rather than to develop an entire system.

From a business and commercial point of view, it is important that the types of system under consideration fit within the terms of an organisation's information technology strategy and, in turn, within the ambit of the more general business strategy that should exist. While enthusiasm for the development of an attractive new system is often understandable, management must ensure that there are sensible business objectives underlying this fervour.

In this context, it is particularly important for management to lay down criteria against which the performance of any system can be measured at specified times in the future. This performance could be in terms of the number of sales of a system, the number of new clients attracted on the strength of the system, the extent of promotional payback, or any other agreed and quantifiable benefit. These performance measurement criteria should be incorporated in the project plan, which in itself will be the guide and touchstone of all involved with the system.

Commercially, it should be borne in mind that technology offers the possibility not just of automating existing practices and procedures but also of innovating, to the extent that IT often allows activities and tasks to be undertaken that were simply not possible prior to the availability of computers. Technology may, therefore, facilitate internally the conduct of Eco-audits, when in the past this would not have been viable. Alternatively, organisations may now perhaps be able to offer Eco-audits more extensively on the strength of the technology-based service.

Concluding remarks

Senior management should recognise a further strategic dimension to the use of IT in Eco-auditing. That is perhaps of greater significance than changes that may result in the conduct of individual audits where technology is used. IT may, in the end, be the most powerful tool to bring about a shift in attitude across industry to environmental regulation. Using technology in the ways described in this chapter will promote a more sensible commercial approach to the law involving less emphasis on legal enforcement and more attention to legal risk management and planning in accordance with the principles introduced throughout this book. Managers should, therefore, be encouraged to think strategically and laterally about the potential of IT as a basis for expanding and developing, and not simply sustaining, existing business practices.

Epilogue: an Update

The European Parliament (EP) met in preliminary session on 19 January 1993 and adopted the Report by Mr Jose Luis Valverde regarding the Eco-audit Regulation. While approving the proposed Regulation, the EP voted in favour of strengthening several of the provisions, as well as expanding the title of the Regulation to a Community Eco Management and Audit Scheme (CEMAS) Regulation. In the words of the report:

> "Changing the title from the 'Community Eco-audit Scheme' to a voluntary 'Community Scheme for Environmental Management Systems' by companies reflects the completeness and the increased usefulness of this scheme in environmental protection."

The Rapporteur has emphasised that this amendment makes it clear that corporate environmental management systems cannot be confined to regular audits but should constitute a whole system of environmental management arrangements, of which audits are merely a part. The scope of the Regulation has, in addition, been extended from just the industrial sector to other activities, including transport and public service activities. The Rapporteur had noted that the inclusion of the service and state end sectors is important in view of their investments in and ownership of industrial companies. In his view the exclusion of retail and distribution firms would damage the credibility, effectiveness and utility of the proposal. The Community should in fact be encouraging companies to participate and, therefore, this encouragement should be Community-wide and should include all business sectors and sizes of enterprises. The Rapporteur had also noticed that the Commission's definition of "site" omitted many transport companies carrying toxic and dangerous substances, even when vehicles were parked. Furthermore, land-based, static "sites" could also be excluded if, for instance, they fell into the retail sector, such as petrol stations or the construction sector, for example temporary sites etc.

Whereas the intention had been that the proposal of 5 March 1992 should be adopted by the Council and be enforced as of 1 January 1993, the Council meeting on 15/16 December 1992 was unable to adopt the proposal, in spite of being very close to agreement, because Germany was opposed to the idea of using a logo. The Germans considered that the use of a logo could cause confusion because if one site benefits, while others do not, the logo would still appear to give some misleading stamp of approval. In addition, the Spanish had a demand for funds to enable small- and medium-sized enterprises (SMEs) to participate in the scheme. There had also been a translation error with regard to the accreditation of auditors; the accreditation system was in fact to be applied only to verifiers not auditors. Accredited environmental verifiers will play a fundamental role in the Eco-audit scheme. They will "verify" a company's environmental policy, programme, management system, review, audit procedure and environmental statement for their conformity with the detailed requirements of the Regulation. In addition, they will "validate" the environmental statement, which is no longer tied to the frequency of the audit. The environmental statement would now be published every year, which means that in the years between audits or completion of audit cycles the statement will be a "simplified" one containing only summary data on a site's releases to the environment and consumption of raw materials, energy and water, and a note of "any significant changes" since the previous statement. Simplified statements will only have to be validated by an environmental verifier at the end of the subsequent audit or audit cycle. Furthermore, where a verifier considers "in particular in the case of small- and medium-size enterprises" that the nature and scale of the operation do not require another statement until the completion of the next audit and where there have been "few significant changes" since the last statement, sites may be exempted from the requirement to prepare a simplified annual statement. Essentially, verifiers will "validate" the environmental statement only where the company has complied fully with the detailed requirements of the Regulation and where the statement itself is accurate, sufficiently detailed and covers all significant environmental issues at the site concerned.

The other amendments concerned the deletion of the reference to health and safety in the Preamble, while the greatest possible participation by workers has now been stressed in the form of a written document.

The EP has also recommended that even though the scheme is voluntary, participation in it should be an indispensable pre-condition for companies wishing to participate in any EC funded programme. Finally, the EP has insisted that the internal audit of the company, and the final verification of a company's environmental statement, should not be carried out by the same person/organisation.

More recently following certain technical amendments the Regulation has been adopted by the Council in an amended version at its meeting on 22 and 23 March 1993. Germany's amendment to the effect that an environmental policy should include commitments aimed at constant environmental improvement in ways corresponding to the economically viable use of "the best technologies available" has been accepted. Spain's insistence on specific provisions for SMEs has also been agreed, provided that support to assist SMEs would not be financial (at Germany's request). Instead of the original logo, the graphic will now consist of a circle of stars as on the Community flag. Within the circle will be included one of three prescribed forms of words applicable to the registered site or participating company. In so far as it is a Regulation it will be binding on Member States in its entirety immediately. Member States will have 21 months to bring the scheme into operation with competent bodies designated within 12 months. Nevertheless, in so far as the scheme is voluntary the effects are likely to bite around July 1993 by which time there should be some indication of the likely level of interest. This additional time should enable companies to look at the costs and benefits of the scheme. As far as the benefits are concerned, it is worth emphasising once more that the openness that is involved in participating in the scheme could help in the avoidance of liability, assisting in planning enquiries and generally in terms of competitive advantage. It is not just a question of achieving a site logo, but a way of truly demonstrating the organisation's awareness of the fact that it is now dealing with a switched on public that are increasingly concerned with improved environmental performance.

In the light of these changes the question of the relationship with environmental standards such as BS 7750 is particularly relevant. The question had been raised whether those certified under the management systems or, for example, the quality system of the chemical industry, would have to go through another process in order to achieve registration under the Eco-audit scheme. It is now clear that those companies which have been certified to national or international standards for environmental management systems will be considered as having met the corresponding requirements of the Regulation on the basis of two conditions. First, the standards concerned should have been formally recognised by the Commission, and the certification must have been carried out by a body whose accreditation has been recognised in the Member State concerned. Secondly, in order to be registered under the EC scheme the companies must meet the requirements of environmental statements. Whether or not BS 7750 will be automatically regarded as equivalent for the purposes of the Eco-audit Regulation is still being debated: the solution will probably be reached only after the pilot studies have been completed later this year. In any event, as seen from earlier discussions, clearly the intention is to attempt to dove-tail the approach at the national, European and international level to enable companies to achieve continuous environmental improvement.

Very recently the Commission has emphasised that the scheme is compatible – even complementary – with initiatives already taken by some private companies such as the Chemical Industry's "Responsible Care Programme". Neither does it, in the Commission's view, encroach on national legislation or interfere with the role of enforcement authorities in the Member States. It is intended to spread a new culture among firms making them take a more proactive position with regard to the environmental effects of their businesses. The Commission considers that because of the checks built into the system it will benefit from credibility vis-à-vis public and national authorities.

It is anticipated that the European Environmental Agency will eventually monitor and supervise the Regulation once the controversy over its location is resolved. In the United Kingdom the Prime Minister, John Major, announced that there would be the creation of a single Environmental Protection Agency (EPA). This would involve the merger of Her Majesty's Inspectorate of Pollution (HMIP) and the National Rivers Authority (NRA). At present it is anticipated that the EPA will come into existence in April 1995 and the relevant Bill is being drafted. This was the indication from Derek Osborn, Director-General, Environmental Protection, DoE, at the Green Alliance Debate on the "Environmental Protection Agency: Hopes and Expectations" held at the Royal Society of Arts on 28 January 1993.

There is a great debate as to the powers and functions of the new Agency. Confusion has already arisen over whether the combined EPA will embark upon a compliance or enforcement route; by way of example, at present HMIP tends towards compliance while NRA adopts a more aggressive enforcement stance over breaches of water regulations. The issue is not made any easier by the powers that are still vested in Local Authorities (*e.g.* Part B list of Integrated Pollution Controls and Waste Regulations) and concerns of commercial businesses that attitudes may vary from one region to another. Although this may not be of direct significance in regard to the voluntary scheme of Eco-auditing, it is an issue that may become more significant with the creation of a European Environmental Protection Agency and the growing commercial importance and legal significance of world standards and management systems.

Unless the position is clarified there is a growing risk that companies will become increasingly uncertain as to how to implement environmental and health and safety audits and systems, the standards which they are required to meet and to whom they are responsible in the event of default. It is in this confused arena that environmental risk protection and particularly confidentiality may become a dangerous minefield for business. However, as the significance of environmental management and auditing grows one anticipates that these difficulties will be overcome and that business will achieve the environmental goals that the public requires and the world needs.

Butterworths EC Brief
Published by: Butterworths, London.

Croner's Europe Bulletin
Published by: Croner, Kingston upon Thames, Surrey.

Croner's Health and Safety Bulletin
Pubished by: Croner, Kingston upon Thames, Surrey.

ENDS
Published by: Environmental Data Services, Camberley, Surrey.

Environment Business
Published by: Information for Industry, London.

Environment Information Bulletin
Published by: Industrial Relations Services, London.

Environment Law Brief
Published by: Legal Studies and Services Publishing Ltd., London.

Environment Law Monthly
Published by: Monitor Press, Great Waldingfield, Sudbury, Suffolk.

Environmental Health
Published by: Institution of Environmental Health Officers.

Environmental Law and Management
Published by: Chancery Law Publishing, London.

Greenpeace Business
Published by: Greenpeace, London.

Greenpeace Campaign Reports
Published by: Greenpeace, London.

Health and Safety Commission Newsletter
Published by: Health and Safety Commission Sheffield.

Journal of Planning and Environment Law
Published by: Sweet & Maxwell, London.

NAWDC News
Published by: National Association of Waste Disposal Contractors, London.

Official Journal of the European Communities
Published by: Office for Official Publications of the European Communities.
Available from HMSO, London.

FURTHER READING

Water Bulletin
Published by: Water Services Association, London.

Water Law
Published by: Chancery Law Publishing, London.

Water and Waste Treatment
Pubished by: Faversham House Group Ltd., Croydon.

Index